Decision Engineering

Navneet Bhushan and Kanwal Rai

Strategic Decision Making

Applying the Analytic Hierarchy Process

With 54 Figures

Springer

Navneet Bhushan, MTech, MSc
Kanwal Rai, MBA, BE
CREAX Information Technologies Pvt. Ltd.
Bangalore, India

British Library Cataloguing in Publication Data

Bhushan, Navneet
Strategic decision making. – (Decision engineering)
1. Decision making 2. Strategic planning 3. Public administration – Decision making
I. Title II. Rai, Kanwal
658.4'032
ISBN 1852337567

Library of Congress Cataloging-in-Publication Data
Bhushan, Navneet, 1966–
Strategic decision making / Navneet Bhushan and Kanwal Rai.
p. cm. -- (Decision engineering)
Includes bibliographical references.
ISBN 1-85233-756-7
1. Decision making--Methodology. 2. Decision making--Mathematical models.
I. Rai,Kanwal, 1973– II. Title. III. Series.
HD30.23.B5 2003 2003059153
658.4'03--dc22

Decision Engineering Series ISSN 1619-5736
ISBN 1-85233-756-7 Springer-Verlag London Berlin Heidelberg
Springer-Verlag is a part of Springer Science+Business Media

Bhushan & Rai: Strategic Decision Making: Applying the Analytic Hierarchy Process

First Indian Reprint 2009

ISBN 978-81-8489-323-6

This edition is manufactured in India for sale only in India, Pakistan, Bangladesh, Nepal and Sri Lanka and any other country as authorized by the publisher.

This edition is published by Springer (India) Private Limited, (a part of Springer Science+Business Media), Registered Office: 3rd Floor, Gandharva Mahavidyalaya, 212, Deen Dayal Upadhyaya Marg, New Delhi – 110 002, India.

Printed in India by Rashtriya Printers, Delhi.

Contents

Preface

Decision making in the dynamic and rapidly evolving world is a major challenge. Decision making essentially involves the generation of a set of alternatives and the choice of the most appropriate alternative for execution by answering the following important questions: what decisions must be made, who will make them, how and what resources will be allocated, and how will the situation will be measured and revisited in the dynamic environment in which the system will be operating. Also, in large organizations such as a multinational business group or a modern nation state, it is imperative to decide what principles, style and guidelines for decision-making are appropriate for the organization. It is essential to decide what structure will govern the process of decision making.

Structured methods utilizing the theoretical and practical advances made in the fields of mathematics, operations research, cybernetics, artificial intelligence, etc, have become an important aid to decision making in all sectors. The theoretical underpinnings of such decision aids is the principle of optimization, which tries to maximize or minimize certain combinations of conflicting variables representing the matrix of interest for the decision maker under constraints imposed by the real life situation. The empirical, common sense or subjective decision making of the past graduated to the field of operations research based on the principle of optimization and has resulted in enhanced decision aids at all levels of an organization.

When the rules of the game are well laid out, when the environment in which one operates is predictable, when the opponents are known, when the actors behave in a deterministic manner, when variables vary within a small and narrow band, and, when linear relations are the norm, one can try to make decisions using the standard optimization techniques. However, when the benefits of actions are unpredictable, when relationships between variables may not only be non-linear and stochastic, but actually unknown, the principle of optimization for decision making will not help much. This is exactly the world that we are facing today. Strategic, operational and tactical agility in quickly responding with maximum concentration of effort is the absolute requirement. At the tactical and operational level standard optimization techniques for decision making have helped to some extent. However, at the strategic levels these techniques have not been able to make a greater impact.

The problems in which stakes are extremely high, human perceptions and judgments are involved and whose solutions have long term repercussions, fall in the strategic level decision-making category. At this level problems are ill defined and are usually in terms that are uncertain, fuzzy and confusing. However, the existing problem-solving techniques based on sound mathematical principles require systematic and well-formed problems. This mismatch between problems and their solution techniques leads to frustration and a lack of confidence by the top decision makers. To solve such problems with limited amounts of time and resources needs the balancing of many variables. This book focuses on applying the Analytic Hierarchy Process (AHP) for such strategic level decision-making problems.

The Analytic Hierarchy Process (AHP) is a systematic approach developed in late 1970s to structure the experience, intuition, and heuristics-based decision making into a well-defined methodology on the basis of sound mathematical principles. The AHP is suited to quantitatively arrive at the decision in the strategic domain. It provides a formalized approach for creating solutions to decision-making problems, where the economic justification of time invested in the decision-making process is reflected in the better quality solutions of the complex decision-making problems.

Strategic level decision making in the three main endeavors of human existence, i.e., Business, Defense and Governance has been described in this book. The book covers a variety of problems in the three domains – from vendor selection to weapon system evaluation, from software projects management to disaster management, from factors affecting national security to factors affecting technology proliferation. Practical case studies from the authors' experiences of many years in applying the AHP in these three domains have been comprehensively dealt with. The range of problems covered in the above three domains of the book gives a comprehensive exposure to the reader to the extent of assistance that a formal methodology such as the Analytic Hierarchy Process (AHP) can provide to a decision maker in evolving strategic decisions in such complex and varied domains in a highly dynamic, uncertain, unknown, and unpredictable world.

Navneet Bhushan and Kanwal Rai

Acknowledgements

This book has covered many years of our practical experience in solving strategic level decision-making problems in multiple domains. During the course of our explorations of this field, a large number of individuals, institutions and clients have influenced our thinking and assisted us in solving these problems either directly or indirectly. We thank all of them. We would however like to mention two names. Mr. Jagjeet Singh Sikka and Dr. S.V. Nagaraj have provided us constant material and moral support during the course of writing this book. Their contribution is acknowledged with gratitude.

I, Navneet Bhushan, would like to dedicate this book to my mother, Mrs. Urmil Satya Bhushan. Over the years, she has been a remarkable source of inspiration and a wonderful guide, besides being a solid pillar of strength. The late Dr. N.K. Jaiswal, who introduced me to the field of AHP, was a brilliant mathematician. This book has been greatly influenced by his work and inputs. I would like to thank Mr. S.C. Jethi, who helped me evolve into an analyst by his keen insights and unparalleled support. Dr. N.K. Jain's support and morale-boosting doses are acknowledged with gratitude. My sisters Kanupriya and Venu Kapoor have always provided unacknowledged support to me. They deserve special mention in this book. Encouragements by Mr. Eshu Jain during this project are acknowledged gratefully. Above all, Ashi Bhushan, my wife, has been more than a co-author of this work. Her love and sacrifices during the course of writing of this book have egged me on and on. Our son Srijan and daughter Snigdha have missed their papa for many hours due to the extra time that I worked to complete this project. I pledge to compensate this loss by investing more time with them from now onwards.

I, Kanwal Rai, can vouch for the fact that writing a book is a journey and is immensely more difficult (at the same time more enjoyable) than mere thinking or planning about it! There have been a number of insights and learnings for me during this journey. It is the immense and irrefutable support from my family that has kept the vigor flowing till the very end. This book is dedicated to my parents who have been a great source of motivation and support – they have always been there to believe in my dreams and me. It is the buoyant energy and inspiration of my wife, Kalpana Sindhu, that has kept me going forward for the seemingly insurmountable. Without her support, conviction and enthusiasm, it would not have been possible to realize this dream. I would also like to extend my thanks to my brother Naveen, sisters Anju and Sumeet, who stood solidly behind me and contributed in their own sweet little ways to make this book a reality.

We owe our thanks to all our colleagues and friends who helped us by providing their precious time to review and criticize the work constructively in order to improve the output. There have been innumerable instances where the feedback has not only helped to improve the quality of content and presentation, but also the authors and their thoughts in person.

PART 1

Strategic Decision-Making and the AHP

1 Strategic Decision-Making

1.1 Introduction

Problems in which the stakes are extremely high, in which human perceptions and judgements are involved, and whose solutions have long-term repercussions, call for a rational approach to their solution. Various techniques are in use for decision-making at the strategic level. However, at this level the problems are ill defined and are usually presented in terms that are uncertain, fuzzy and confusing, while problem-solving techniques based on sound mathematical principles can only be applied to systematic and well-formed problems. This mismatch between problems and their solution methods leads to the frustration of top decision-makers and their loss of confidence in mathematical techniques.

To solve such problems using reasonable amounts of time and resources requires the juggling of many variables. The focus should be on developing a comprehensive methodology for solving strategic-level decision-making problems which are at present tackled in an ad-hoc manner.

In today's highly uncertain world making a decision which has long-term implications requires a thorough understanding of likely or possible future situations and also the ability to balance a large number of controllable and uncontrollable parameters. However, the time now given to decision-makers to reach high-risk, long-term decisions is decreasing. The world becomes ever more unstable, more disordered and more uncertain and hence requires more and more and better and better analytical tools for making such decisions. Therefore, a thorough, systematic framework based on a scientific footing is needed to analyse and make appropriate decisions in the world as it is today.

The strategic decision-making process can be described as shown in Figure 1.1. Every nation or company in the world exists along with other nations or companies in some group or other; none can exist in isolation. And different nations and companies have different national or market interests that may clash. This leads each nation or company to feel threatened by other nations or companies, and, in turn, a threat perceived calls for methods to assess it. Decision-makers (DM) need to find out the extent of the threat or competition which their nation or company faces from the others. Once the threat is assessed, strategies or courses of action (COA) are generated to meet it and to achieve strategic objectives within the framework of execution doctrine. A methodology with a scientific basis is also required in order to determine alternative strategies. Once a large number of strategies have evolved,

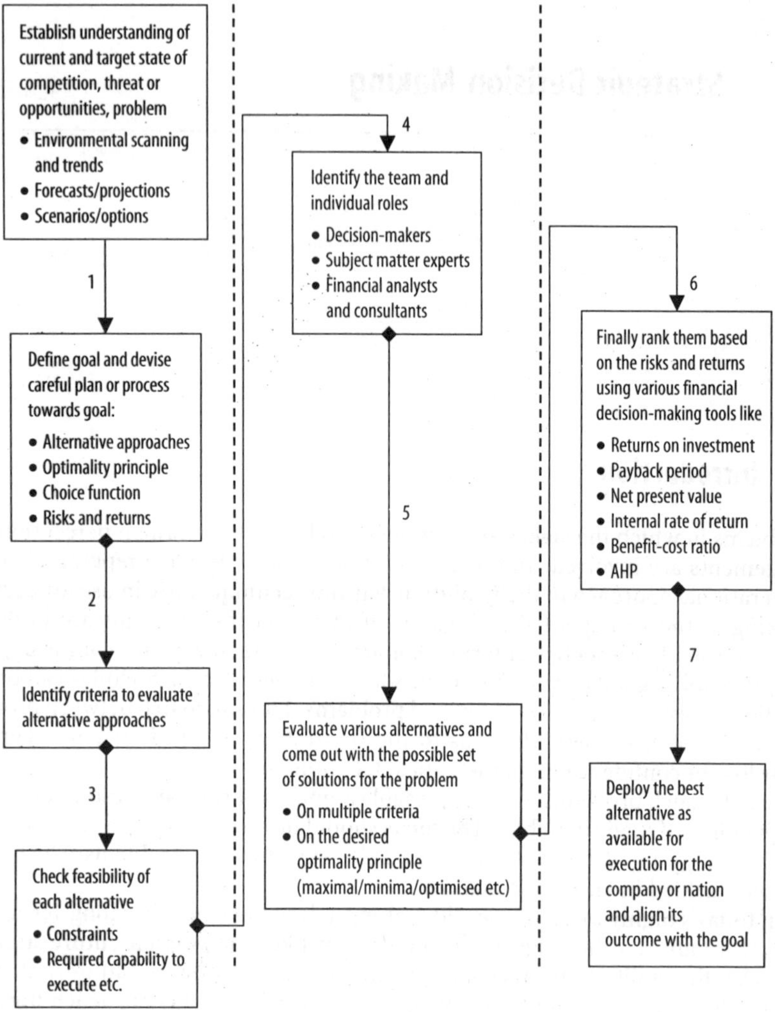

Figure 1.1 Strategic decision-making process.

these have to be evaluated in terms of the likely future scenario and the best is chosen and executed. To choose those strategies, which can cater for future crisis and contingency situations, the DM needs guidance or at least some form of mechanism.

1.2 Strategic Decision-Making Process

Competition and threat drive businesses and economies to make rapid choices amid a dynamically changing pattern of chaos. Those choices that are strategic critically influence success or failure in the long term.

Strategy involves "fit". An organisation's internal capabilities must fit the external environment in which it operates. The internal-external paradigm is the basis for many strategic planning models, with always the underlying assumption being that the internal factors are controllable and the external ones uncontrollable.

Decision-making involves "choice". For achieving the objective or goal, an executive or DM may generally have more than one alternative at hand (*choice function: set of choices*). Usually the DM selects the best alternative based upon his experience, intuition and judgement. This leads to qualitative and subjective decision-making, which may or may not be optimal (*optimality principle: set of criteria to be minimised or maximised and constraints to be overcome*).

Strategic decision-making (SDM) involves fitting the internal capabilities to the external environment by choosing the best among the possible alternatives. Scientific analysis helps in evaluating various alternatives quantitatively and providing the DM with a rational basis for selecting the optimal option.

The critical problems associated with the SDM process are:

1. *Uncertainty:* Coping with uncertainty resulting from inadequate knowledge and excessive complexity.
2. *Self-fulfilling and self-defeating prophecies*: Coping with the fact that conditions aren't fixed externally but are strongly affected by decisions.
3. *Fragmentation*: Coping with the fragmentation of the policy-planning process into isolated but connected regional functional groups.

The standard approach employed in the strategic planning process is to take a set of fixed interests, put them with a fixed environment and then, given the constraints imposed by the environment, invent a strategy for attaining one's interests. In short-range planning such a model is useful. In the long run, however, uncertainty erodes the foundations of this model. The interests should be stated at a level of generality high enough for them not to change greatly during the period under consideration.

To cope up with the uncertainty in the environment, multiple alternative future environments should be sufficiently few in number to be intellectually manageable, but numerous enough to display most of the important alternative outcomes of current world trends.

In view of the SDM process described in Figure 1.1, we can say in summary that a framework for SDM should have the following features:

1. A methodology or model to forecast future situations or scenarios. Scenario writing demands not only an understanding of today's realities but also imaginative forays into tomorrow's possibilities.
2. An environment for evolving strategies to take care of future situations.
3. Techniques to assess situations and understand the behaviour of the system.
4. Techniques to incorporate expert opinions into the SDM process.
5. Methodology to choose the *best* among the alternative strategies.

Coming up with new strategies to achieve a particular objective involves a high degree of creativity and balanced judgement. It requires an innate ability to look into the future, with remarkable farsightedness and the ability to balance various parameters which may change in times to come. Understanding the future and taking account of every parameter in the uncertainty that lies ahead, so as to come up with a rational strategy for achieving an objective, is a difficult task. This task cannot be carried out in an ad-hoc manner based only on somebody's experience, intuition and judgement, especially if the stakes of a particular choice of strategy are extremely high and the wrong or poor choice may lead later to disastrous effects.

Before they come up with various alternative strategies, decision-makers must foresee possible contingencies or crisis situations. A crisis is defined as a crucial stage or turning point, which implies a fairly compressed period of time.

As already stated, to come up with possible strategies for coping with various crisis situations, the DM requires a framework with a scientific footing. This demands a thorough understanding of possible crisis situations, how these may be avoided and how to manage them if they occur. This means that strategies for managing any future crisis situations should be outlined well in advance, while one should strive to create strategies that will avoid crisis situations as far as possible. In other words, one needs to work out strategies both for crisis avoidance and for crisis management (see Chapter 9).

Once the strategies have evolved their performance in the likely future environment needs to be evaluated. This in itself becomes an important activity, with four methods generally used, namely expert opinion, direct measurement, analytical modelling and system simulation. Usually the direct measurement of the performance in the operational environment is prohibited by cost. Analytical modelling becomes extremely tedious and if simplified becomes unrealistic.

Expert opinions about the performance of a strategy in the likely scenarios are normally vague, confusing and fuzzy. However, expert judgements are important and can be used if they are quantified (see Chapter 7).

1.3 Commonly Used Decision Analysis Tools

Various decision analysis tools are deployed in the decision-making process, each with its strengths and weaknesses. It is up to the decision-maker or makers to understand the context, underlying advantages and disadvantages of these tools prior to their deployment in the decision-making process.

1.3.1 Net Present Value (NPV)

This measures the present worth of the multi-year investments. All the cash transactions (inflows and outflows) are discounted to the same point of time (typically the beginning of the project).

The project having the largest positive NPV is ranked as the best.

Strengths: NPV is a very simple method to use and is especially helpful in evaluating long-term projects. The NPV method considers the time value of money and allows the use of varying discount rates.

Weaknesses: NPV is highly sensitive to the discount percentage and computing the percentage can be difficult and controversial. NPV alone is not a good criterion for decision-making as it doesn't consider the absolute value of the investment and hence the risk associated with the size of the investment.

1.3.2 Internal Rate of Return (IRR)

This is the discount rate at which the NPV of cash flows becomes zero. It is used as a cutoff rate for investment decisions – the investment is acceptable if the IRR is greater than the opportunity cost of capital. The project with the highest IRR could be the best investment.

Strengths: The IRR allows comparison of investments of different sizes across different businesses. Unlike NPV, it does not require calculation of discount rate.

Weaknesses: The unrealistic assumption that profits can be reinvested at the same IRR could produce misleading results. Unlike the NPV, the IRR is not easy to compute, as it involves an iterative process.

1.3.3 Benefit–Cost Ratio (BCR)

This compares the total cost of implementing an initiative with the total monetary benefits derived after implementation as:

$$\text{BCR} = \frac{\text{Total benefits}}{\text{Total costs}}$$

An investment option is considered to be rational if it has a BCR above 1. An investment option with largest value of BCR (over and above 1) is considered to be the best.

Strengths: Commonly used for make-or-buy decisions, BCR is a repeatable and objective test of the profitability of an investment option at the highest level when not much finer details are charted out.

Weaknesses: The monetarization of intangible benefits is a subjective activity. One can maximise the benefits while omitting the inclusion of probable cost factors in the total costs, and thereby skew the BCR, allowing manipulation in favour of the decision maker.

1.3.4 Total Cost of Ownership (TCO)

This is a method for identifying all the costs associated with the investment option across its life cycle until it remains effective for an organisation. For example, owning an asset includes costs under various categories associated with its lifecycle such as acquisition, maintenance, training, customisation etc.

Strengths: TCO helps to identify the stream of costs across the useful life of the subject of the investment and to strengthen the cost analysis.

Weaknesses: TCO doesn't capture the financial benefits of the investment project. TCO alone is not sufficient for decision-making.

1.3.5 Payback Period

The payback period for an investment is the period when the cumulative cash flow equals the initial investment. Therefore, the payback period is the time taken to earn back the initial investment. Projects with shorter payback periods are preferred.

Strengths: It's easy to understand and compute. It provides a good indicator for differentiating risks by separating long-term projects from short-term ones.

Weaknesses: The method does not consider all the cash flows after the break-even period and therefore is not able to provide a true picture of the profitability of a project. It also doesn't take into consideration the time value of money.

1.3.6 Balanced Scorecard

A prescriptive framework for measuring an organisation's performance in four key areas: financial, customer, learning and internal processes. It helps an organisation translate strategy into objectives, measurements, targets and initiatives (see Chapter 3).

Strengths: It is a holistic approach, which measures not only financial but also non-financial performance for investment decisions. It helps to evaluate an investment's impact on the entire business unit or company. The integrated scorecard approach details out metrics across the different management layers.

Weaknesses: Defining and maintaining a balanced scorecard is a very cumbersome and time-consuming activity. This could result in the diversion of precious resources and management time in irrelevant metrics. The approach does not consider the relative importance of the different metrics it uses.

1.3.7 Economic Value Added (EVA)

This measures the incremental value created by an investment and is measured by:

$$\text{Return on capital} - \text{Cost of capital} \times \text{Capital invested in the project}$$

Strengths: EVA prevents managers from assuming that the cost of capital is free. It is a good economic decision-making tool that helps enterprises to relate, maximise and communicate the results of valuation exercises.

Weaknesses: EVA uses the discounted cash flow (DCF) technique and suffers the same inherent drawbacks, i.e. the computation of the cost of capital could be controversial and manipulative.

1.3.8 Return on Investment (ROI)

This is a generic term known by its different forms such as "return on assets", "return on equity", "return on capital employed" etc. The underlying intention is to calculate: (Gains − Investments) ÷ Investments. The results are obtained in percentages. The project with the highest ROI is considered to be the most lucrative for investment purposes.

Strengths: ROI exists in multiple forms that help apples-with-apples comparison of returns from investments (by category). It can be a good tool for comparing short-term projects with similar patterns of returns.

Weaknesses: ROI doesn't account for the risk and magnitude of the investment being made. It ignores the opportunity cost of capital. The inclusion of intangible gains by quantifying them could manipulate the ROI and can skew the decision in favour of manipulation. Changing the time period of investment and returns can dramatically change ROI estimates.

1.4 Characteristics of a Formal Strategic Decision-Making Framework

The standard approach employed in the strategic planning process is to take a set of fixed interests, put them with a fixed environment and then invent a strategy for attaining one's interests given the constraints imposed by the environment. In a short-range planning process such a model is useful. In the long run, however, uncertainty erodes the foundations of this model, impacting on interests, environment and strategy. Thus interests should be stated at a level of generality high enough for them not to change greatly during the period under consideration. To cope with uncertainty in the environment, multiple alternative future environments are described. These should be few enough in number to be intellectually manageable, but numerous enough to display most of the important alternative outcomes of current world trends. A mistake policy-makers normally make at this juncture is to resort to contingency planning, with an alternative strategy described for each alternative future. However, long-range planning seeks to facilitate decision-making in the present, and present decisions can follow only one strategy, not several.

Typically in strategic planning, the environment is demarcated into a *core environment*, which will remain more or less static during the planning period, and alternative perceived environments forecast by experts. Besides these, certain exogenous contingencies should also be considered. To cope with the core environment a *core strategy* is needed, supplemented by a *basic strategy* whose dual purpose is both to influence the environment towards the optimal one and to facilitate *success* within that environment. The core strategy deals with the constants of the environment, whereas the basic strategy deals with the variable features. A *hedging strategy* is also needed to counter unforeseen contingencies. It is required because the alternative environments chosen for study can never completely cover the range of possible alternatives; the basic strategy may fail to change the world and random events may occur in the environment.

In view of the above, a framework for SDM should have the following features:

- A methodology or model to forecast future world situations.
- An environment for evolving strategies (core, basic and hedging strategies) to take care of future situations.
- Techniques to assess situations and understand system behaviour.
- Techniques to incorporate expert opinions into the SDM process.
- Methodology to choose the *best* strategy from among the alternative strategies.

A framework for strategic decision-making possessing these features requires a common language for communication between all stakeholders. It needs a formal process or method that can help crystallise the thoughts for the specific problem at hand and provide a common basis for all decision-makers to understand and contribute in the process together. The analytic hierarchy process (AHP), a formal decision-making methodology developed in the late 1970s, is a very attractive basis for forming such a framework.

References

1. Nande P, Schotz E (1988)A hybrid model of strategic planning, Improving decision making in organisation. In: Lockett AG, Islei G (eds) Lecture Notes in Economics and Mathematical Systems 335, Springer.
2. Shubik M (ed.) (1991) Risk, Organisations and Society. Kluwer, Dordrecht.
3. Bhushan N (1991) Risk, Organisations and Society – book review. Opsearch, Journal of the Operational Research Society of India 28(4): 308.
4. Clark DN, Scott JL (1995) Strategic level MS/OR tool usage in the United Kingdom: An empirical survey. Journal of the Operational Research Society (UK) 46(9): 1041–1051.
5. Jaiswal NK (1997) Military Operations Research: Quantitative Decision-making. Kluwer, Dordrecht.
6. Mood AM (1983) Introduction to Policy Analysis. Elsevier.
7. Frci D, Ruloff D (1989) Handbook of Foreign Policy Analysis. Nijhoff.
8. Miser HJ, Quade ES (1988) Handbook of Systems Analysis. Elsevier, London.
9. Asher W, Overholt WH (1983) Strategic Planning and Forecasting. Wiley, New York.

2 The Analytic Hierarchy Process

2.1 Do You Need a Formal Decision-Making Framework?

The complexity of the modern world is a much-acknowledged fact. As the human race develops, complexity increases. Technology has created various artefacts to relieve us of manual, routine and time-consuming tasks. The predictable and deterministic world of the past has been replaced by the uncertain, random and disorderly world of today. Technological advances in multiple fields of human activity have created a planet on which things happen at electronic speed. Rapidly increasing complexity and information overload have schemed together to drastically reduce the time available for making decisions. The decision-maker is stressed, overloaded with unsolicited information, has not enough time to analyse the situation, and yet must make decisions that have high-risk implications or consequences. What does the decision-maker need? Human decision-making in the world characterised above needs a quick-response analysis of the situation that some how captures the decision-maker's intuition, judgement and experience. This can then be combined with detailed quantitative analysis based on the information glut that is churned out from the plethora of process measurements, balanced scorecards, business intelligence, data accumulation and information generation techniques and systems in place in various organisations.

Decision-making, especially strategic decision-making, with high stakes and stochastic future implications, involves multiple actors. In most organisations, these decisions are made collectively, irrespective of whether the organisation is a privately owned business, a public limited company or a government agency. This is true for national, international and multinational organisations as well. Even in small and medium enterprises decision-making is rarely done by a lone individual sitting in isolation. The reality of a group making these high-stakes decisions generates a requirement for creating communication links between the members of the decision-making group with a common understanding of the syntax and semantics of the underlying issues. Decisions made in an ad-hoc, unstructured or semi-structured manner, based on the availability of only a subset of the decision-making group at the time of decisions, has a high probability of being not just sub-optimal but utterly wrong, with disastrous results.

The single-criterion and simple decision-making requirements of the past have today given way to highly complex decision problems involving multitudes of variables, which may be stochastic, fuzzy or at worst unknown. As the time required to make decisions has been severely reduced, the onus of decision-making has shifted to the lowest level of the hierarchy of organizations.

Thus we can infer that hierarchical organizations need a comprehensive formal framework for decision-making owing to increasing complexity and stochasticity, the involvement of many decision-makers and the shift in decision-making requirements to field-level workers, as explained above. It is not that this has become a sudden requirement; in the past formal decision analysis techniques were developed to tackle these problems. However, these have been found to be too mathematical or theoretical or else capable only of solving older problems totally different from those of today.

Structured methods utilising the theoretical and practical advances made in the fields of mathematics, operations research, cybernetics, artificial intelligence etc. have become important aids to decision-making in all sectors. The theoretical underpinnings in such decision aids is the principle of optimisation, which tries to maximise or minimise certain combinations of conflicting variables which represent the metric of interest for the decision-maker under constraints imposed on these variables by the real-life situation. This principle has resulted in an enormous intellectual expansion of quantitative decision-making aids using standard optimisation techniques. Empirical, common-sense or subjective decision-making, supplemented by some simple calculations using arithmetic, geometry and calculus, has evolved into techniques of sophisticated operations research based on the principle of optimisation and has resulted in enhanced decision aids at every level of organisation, thanks to increasing automation in the form of the computerisation of the techniques involved.

2.2 Formal Decision-Making Techniques

When the rules of the game are well laid out, when the environment in which one operates is predictable, when the opposition is known, when the actors behave in a deterministic manner, when costs vary within a small, narrow band, and, when linear relations are the norm, one can try to make decisions using the standard optimisation techniques. However, when the benefits of actions are unpredictable, when relationships between variables may be not only non-linear and stochastic but also actually unknown, the principle of optimisation for decision-making will not help much. This is exactly the situation we face in the world of today. Strategic, operational and tactical agility, in quickly absorbing a situation and responding with maximum concentration of effort at the point of need, is the absolute requirement. At the tactical and operational level in various large-scale organisations, standard optimisation techniques for decision-making have in the more orderly world of the past helped to some extent. However, at the strategic level these techniques have been unable to make any greater impact.

Decision-making can be considered as the choice, on some basis or criteria, of one alternative among a set of alternatives. A decision may need to be taken on the basis of multiple criteria rather than a single criterion. This requires the assessment of various criteria and the evaluation of alternatives on the basis of each criterion and then the aggregation of these evaluations to achieve the relative ranking of the alternatives with respect to the problem. The problem is further compounded when there are several or more experts whose opinions need to be incorporated in the decision-making. It is lack of adequate quantitative information which leads to dependence on the intuition, experience and judgement of knowledgeable persons called experts.

We can define a generic decision-making problem as consisting of the following activities:

- Studying the situation.
- Organising multiple criteria.
- Assessing multiple criteria.
- Evaluating alternatives on the basis of the assessed criteria.
- Ranking the alternatives.
- Incorporating the judgements of multiple experts.

The problem can be abstracted as how to derive weights, rankings or importance for a set of activities according to their impact on the situation and the objective of decisions to be made. This is the process of multiple-criteria decision-making (MCDM). The MCDM problems have been studied under the general classification of operations research (OR) problems, which deal with decision-making in the presence of a number of often conflicting criteria. The field of MCDM is divided into multi-objective decision-making (MODM) and multi-attribute decision-making (MADM). When the decision space is continuous, MODM techniques such as mathematical programming problems with multiple objective functions are used. On the other hand, MADM deals with discrete decision spaces where the decision alternatives are predetermined. Many of the MADM methods have a common notion of alternatives and attributes. Alternatives represent different choices of action available to the decision-maker, the choice of alternatives usually being assumed to be finite. Alternatives need to be studied, analysed and prioritised with respect to the multiple attributes with which the MADM problems are associated. Attributes are also referred to as goals or decision criteria. Different attributes represent different dimensions of looking at the alternatives, and may be in conflict with each other, may not be easily represented on a quantitative scale – and hence may not be directly measurable – and may be stochastic or fuzzy. Further, these attributes may have totally different scales – quantitative or qualitative. Most of the MADM methods require that each attribute is given a weight or relative importance with respect to their impact on the decision problem being solved. MADM and MCDM have often been used to mean the same class of models; here we will use the more commonly used term MCDM to denote MADM problems.

The weighted-sum method (WSM), or the decision matrix approach, is perhaps the earliest method employed. This evaluates each alternative with respect to each criterion and then multiples that evaluation by the importance of the criterion. This product is summed over all the criteria for the particular alternative to generate the rank of the alternative. Mathematically,

$$R_i = \sum_{j=1}^{N} a_{ij} w_j \tag{2.1}$$

where R_i is the rank of the ith alternative, a_{ij} is the actual value of the ith alternative in terms of the jth criterion, and w_j is the weight or importance of the jth criterion.

Let us assume there are two criteria, C1 and C2, and three alternatives, A1, A2 and A3. Let us assume that the weights assigned to the criteria C1 and C2 are $W_1 = 20$

Table 2.1 Decision matrix approach.

Alternative	*Criterion*		*Weighted sum* $R_i = \sum_{j=1}^{N} a_{ij}w_j$
	C1 (W_1 = *20*)	*C2* (W_2 = *30*)	
A1	$a_{11} = 5$	$a_{12} = 5$	250
A2	$a_{21} = 7$	$a_{22} = 3$	230
A3	$a_{31} = 11$	$a_{32} = 3$	310

Table 2.2 Forced decision matrix.

Alternative	*A1*	*A2*	*A3*	*Score (S)*	*Rating* = *S/N*
A1		1	1	2	= 2/3 = 0.67
A2	0		1	1	= 1/3 = 0.33
A3	0	0		0	= 0/3 = 0.00
Total number of comparisons (*N*) =				3	

and $W_2 = 30$, respectively. Each of the alternatives is evaluated with respect to each criterion. The computations are shown in Table 2.1.

Subjectivity, bias and prejudice in giving these ratings and weights cannot be eliminated or evaluated in this method. The additive utility assumption on which this method is based creates problems when the units of the multiple criteria differ from one other.

A variant of the decision matrix approach is the forced decision matrix (FDM) approach. In this, ratings are given in terms of 0 or 1. This winner-takes-all approach is easier to implement, because if a particular alternative is better on one parameter then the whole weight of that parameter goes to the alternative. The FDM approach is illustrated in Table 2.2. The table shows the evaluation of three alternatives, A1, A2 and A3, with respect to single criteria. In the FDM, pairwise comparisons of alternatives are made. As there are three alternatives we need to make three pairwise comparisons so that each alternative is compared with the others once.

The weighted-product method (WPM) is very similar to the weighted-sum method (it also is called dimensionless analysis). Each alternative is compared with others by multiplying a number of ratios, one for each criterion. Mathematically, the comparison of alternatives A1 and A2 will be done as given in Equation (2.2).

$$R(A_1/A_2) = \prod_{j=1}^{N} (a_{1j}/a_{2j})^{w_j} \tag{2.2}$$

where N is the number of criteria, a_{ij} is the actual value of the ith alternative in terms of the jth criterion and w_j is the weight of the jth criterion.

Two other methods, namely ELECTRE (elimination and choice translating reality) and TOPSIS (technique for order preference by similarity to ideal solution), have been described in the literature. The ELECTRE method is similar to the FDM method in principle as it deals with outranking relations by using pairwise comparisons. The basic concept of the TOPSIS method is that the selected alternative should have the shortest distance from the ideal solution and the furthest distance from the negative-ideal solution in a geometrical sense. Here we will not be discussing these two methods, as they are beyond the scope of the present discussion. The reader is referred to [1, 7] for details on these methods.

The analytic hierarchy process (AHP) is a systematic approach developed in the 1970s to give decision-making based on experience, intuition and heuristics the structure of a well-defined methodology derived from sound mathematical principles. It provides a formalised approach where economic justification of the time invested in the decision-making process is provided by the better quality of the solutions to complex problems.

2.3 The Analytic Hierarchy Process – Background

The AHP is based on the experience gained by its developer, T.L. Saaty, while directing research projects in the US Arms Control and Disarmament Agency. It was developed as a reaction to the finding that there is a miserable lack of common, easily understood and easy-to-implement methodology to enable the taking of complex decisions. Since then, the simplicity and power of the AHP has led to its widespread use across multiple domains in every part of the world. The AHP has found use in business, government, social studies, R&D, defence and other domains involving decisions in which choice, prioritization or forecasting is needed.

Owing to its simplicity and ease of use, the AHP has found ready acceptance by busy managers and decision-makers. It helps structure the decision-maker's thoughts and can help in organizing the problem in a manner that is simple to follow and analyse. Broad areas in which the AHP has been applied include alternative selection, resource allocation, forecasting, business process re-engineering, quality function deployment, balanced scorecard, benchmarking, public policy decisions, healthcare, and many more. Basically the AHP helps in structuring the complexity, measurement and synthesis of rankings. These features make it suitable for a wide variety of applications. The AHP has proved a theoretically sound and market-tested and accepted methodology. Its almost universal adoption as a new paradigm for decision-making coupled with its ease of implementation and understanding constitute its success. More than that, it has proved to be a methodology capable of producing results that agree with perceptions and expectations.

2.4 The AHP – Step by Step

The AHP provides a means of decomposing the problem into a hierarchy of subproblems which can more easily be comprehended and subjectively evaluated. The subjective evaluations are converted into numerical values and processed to rank each alternative on a numerical scale. The methodology of the AHP can be explained in following steps:

Step 1: The problem is decomposed into a hierarchy of goal, criteria, sub-criteria and alternatives. This is the most creative and important part of decision-making. Structuring the decision problem as a hierarchy is fundamental to the process of

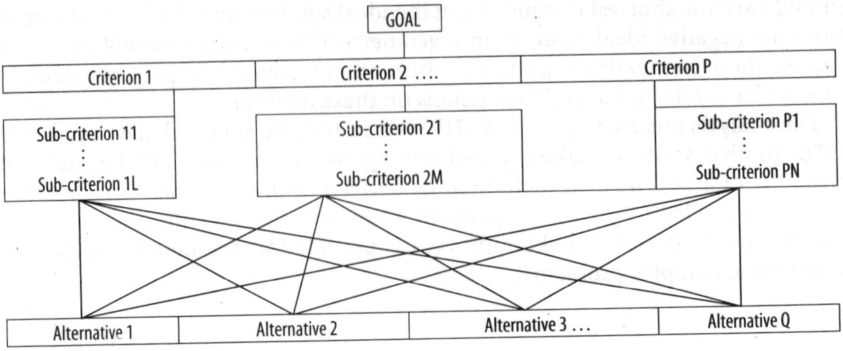

Figure 2.1 Generic hierarchic structure.

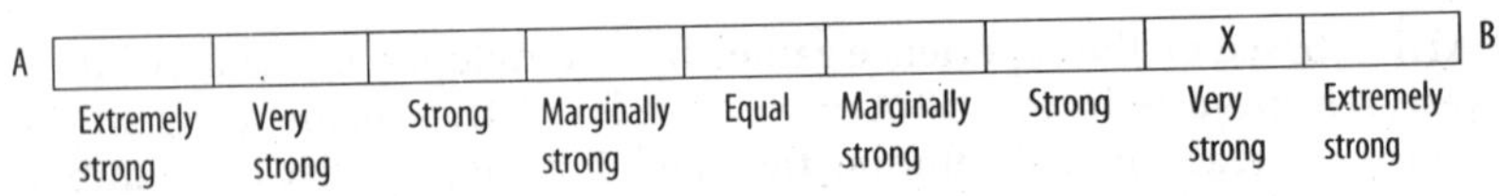

Figure 2.2 Format for pairwise comparisons.

the AHP. Hierarchy indicates a relationship between elements of one level with those of the level immediately below. This relationship percolates down to the lowest levels of the hierarchy and in this manner every element is connected to every other one, at least in an indirect manner. A hierarchy is a more orderly form of a network. An inverted tree structure is similar to a hierarchy. Saaty suggests that a useful way to structure the hierarchy is to work down from the goal as far as one can and then work up from the alternatives until the levels of the two processes are linked in such a way as to make comparisons possible. Figure 2.1 shows a generic hierarchic structure. At the root of the hierarchy is the goal or objective of the problem being studied and analysed. The leaf nodes are the alternatives to be compared. In between these two levels are various criteria and sub-criteria. It is important to note that when comparing elements at each level a decision-maker has just to compare with respect to the contribution of the lower-level elements to the upper-level one. This local concentration of the decision-maker on only part of the whole problem is a powerful feature of the AHP.

Step 2: Data are collected from experts or decision-makers corresponding to the hierarchic structure, in the pairwise comparison of alternatives on a qualitative scale as described below. Experts can rate the comparison as equal, marginally strong, strong, very strong, and extremely strong. The opinion can be collected in a specially designed format as shown in Figure 2.2.

"X" in the column marked "Very strong" indicates that B is very strong compared with A in terms of the criterion on which the comparison is being made. The comparisons are made for each criterion and converted into quantitative numbers as per Table 2.3.

Step 3: The pairwise comparisons of various criteria generated at step 2 are organised into a square matrix. The diagonal elements of the matrix are 1. The criterion in the ith row is better than criterion in the jth column if the value of element (i, j) is more than 1; otherwise the criterion in the jth column is better than that in the ith row. The (j, i) element of the matrix is the reciprocal of the (i, j) element.

Table 2.3 Gradation scale for quantitative comparison of alternatives.

Option	*Numerical value(s)*
Equal	1
Marginally strong	3
Strong	5
Very strong	7
Extremely strong	9
Intermediate values to reflect fuzzy inputs	2, 4, 6, 8
Reflecting dominance of second alternative compared with the first	Reciprocals

Step 4: The principal eigenvalue and the corresponding normalised right eigenvector of the comparison matrix give the relative importance of the various criteria being compared. The elements of the normalised eigenvector are termed weights with respect to the criteria or sub-criteria and ratings with respect to the alternatives.

Step 5: The consistency of the matrix of order n is evaluated. Comparisons made by this method are subjective and the AHP tolerates inconsistency through the amount of redundancy in the approach. If this consistency index fails to reach a required level then answers to comparisons may be re-examined. The consistency index, CI, is calculated as

$$\mathrm{CI} = (\lambda_{max} - n)/(n - 1)$$

where λ_{max} is the maximum eigenvalue of the judgement matrix. This CI can be compared with that of a random matrix, RI. The ratio derived, CI/RI, is termed the consistency ratio, CR. Saaty suggests the value of CR should be less than 0.1.

Step 6: The rating of each alternative is multiplied by the weights of the sub-criteria and aggregated to get local ratings with respect to each criterion. The local ratings are then multiplied by the weights of the criteria and aggregated to get global ratings.

The AHP produces weight values for each alternative based on the judged importance of one alternative over another with respect to a common criterion.

2.5 AHP – Theory

Saaty [19] describes the seven pillars of the AHP as follows:

- Ratio scales, proportionality and normalised ratio scales.
- Reciprocal paired comparisons.
- The sensitivity of the principal right eigenvector.
- Clustering and using pivots to extend the scale.
- Synthesis to create a one-dimensional ratio scale for representing the overall outcome.
- Rank preservation and reversal.
- Integrating group judgements.

The use of ratio scales for comparisons helps in unifying the multidimensionality of the problem in a unified dimension from the perspective of the final outcome. Comparison of oranges and apples can be achieved if their properties are reduced to dimensionless quantities such as the ratios of the properties in some specific dimension or measurement. Ratios are invariant under multiplication by a positive quantity.

For example, if a steel rod of length A metres is compared with a wooden rod of length B metres, it is easy to ascertain the difference between these two rods, since they are measured in the units. The length of the steel rod is $A - B$ metres more or less than that of the wooden rod, depending upon whether $A - B$ is a positive or a negative quantity. Now let us assume that the weight of the steel rod is U kilograms and that of the wooden rod V kilograms. Thus we can compare the weights as $U - V$, finding that the steel rod is $U - V$ kilograms heavier or lighter than the wooden one. We can see that measurement of two unique properties is quite possible using some specific units. Now look at the problem of comparing two properties. The question asked is how one would compare the two rods in terms of length and weight. The traditional answer will be that the difference in length is $A - B$ metres and difference in weight is $U - V$ kilograms. Which one should the decision-maker choose? It depends upon the importance that the decision-maker gives to weight or length. Let us assume Imp1 is the importance given to weight, Imp2 that given to length. Can the decision-maker then choose the steel or the wooden rod based on the quantity Imp1 $\times$ $(A - B)$ + Imp2 $\times$ $(U - V)$? The answer is obviously no, as the units do not match.

The trick lies in eliminating the units. Instead of difference, if we had taken the ratio of lengths and ratio of weights of the two rods, we could easily have compared the two rods with respect to multiple dimensions. This implies that if we take the ratio of steel rod length and wooden rod length, i.e. A/B, and the ratio between the weights of the two rods, i.e. U/V, we can easily choose the rod which depending upon the quantity Q = Imp1 $\times$ A/B + Imp2 $\times$ U/V. Let us give values to these variables, say $A = 20$ metres, $B = 80$ metres, $U = 50$ kg, $V = 25$ kg, Imp1 = 20 and Imp2 = 10; then $Q = 20 \times (20/80) + 10 \times 50/25 = 25$. What do we do with this Q? Q has to be put in perspective that total importance of both these dimensions, i.e. Imp1 + Imp2 = 20 + 10 = 30. Hence when we get $Q = 25$ it has to be compared with 30. The ratio comes out to be less than 1 (i.e. 25/30), hence the decision-maker can choose the wooden rod. Another way of doing this is to normalise the importance of the criteria; if Imp1 = 20/30 = 0.67 and Imp2 = 10/30 = 0.33, it will help in making the computations simple. How do we get Imp1 and Imp2? We can take the ratios again!

It is perhaps easy to measure the lengths and weights of two distinct objects in some predetermined units such as metres and kilograms and then compare them by taking the ratio between the measured quantities. However, when asked to compare two objects or persons with respect to abstract properties such as beauty, honesty, smartness, etc., how does one do it? In this scenario, units for absolute measurement are missing. Not only that: absolute measurement of the two distinct objects being compared is actually not needed. It is the relative measurement that is the essence of comparison. This fact, that only relative measurement is needed, is the fundamental pillar of the AHP. Once we realise that only relative measurement is needed, it means that, at a particular point in time, we need to compare only two objects with respect to the property, criterion, sub-criterion or goal as the case may be. This realisation leads us to paired comparisons. We have now reached the conclusion that relative, paired comparisons are what decision-makers actually do or should do. Since we have taken a ratio of two objects with respect to an attribute, it is easy to translate it into a reciprocal relationship, i.e. if A compares w_1/w_2 times compared with B then B compares w_2/w_1 times compared with A. *Reciprocal, paired comparisons for relative measurement are the second pillar of the AHP.* The measurement scale defined for the AHP is one of 1 – 9 in absolute numbers.

If A is a consistent matrix, small perturbations in A do not lead to perturbations in the principal eigenvector of A. If the order of the matrix, n, is small then small perturbations in A do not create perturbations in the principal eigenvector. The AHP allows for clustering to extend the comparison scale from $1-9$ to $1-\infty$. Taking an alternative in a cluster with properties measured in the same order and comparing it with higher-order alternatives can perform this function. In this way, very small alternatives can be compared with very large ones. The synthesis of global priorities at each level of the hierarchy is carried out by a multilinear form of elements of priority vectors at the lower levels. The AHP has a well-established theory and guidelines for when to preserve rank and when to allow it to reverse. The AHP also provides a methodology to allow the aggregation of individual judgements for taking group decisions.

Theoretically the AHP is based on four axioms given by Saaty; these are:

Axiom 1: The decision-maker can provide paired comparisons a_{ij} of two alternatives i and j corresponding to a criterion/sub-criterion on a ratio scale which is reciprocal, i.e. $a_{ji} = 1/a_{ij}$.

Axiom 2: The decision-maker never judges one alternative to be infinitely better than another corresponding to a criterion, i.e. $a_{ij} \neq \infty$.

Axiom 3: The decision problem can be formulated as a hierarchy.

Axiom 4: All criteria/sub-criteria which have some impact on the given problem, and all the relevant alternatives, are represented in the hierarchy in one go.

In nutshell, there are three major concepts behind the AHP, as follows:

The AHP is analytic – mathematical and logical reasoning for arriving at the decision is the strength of the AHP. It helps in analysing the decision problem on a logical footing and assists in converting decision-makers' intuition and gut feelings into numbers which can be openly questioned by others and can also be explained to others.

The AHP structures the problem as a hierarchy – Hierarchic decomposition comes naturally to human beings. Reducing the complex problem into sub-problems to be tackled one at a time is the fundamental way that human decision-makers have worked. Evidence from psychological studies suggests that human beings can compare 7 ± 2 things at a time. Hence to deal with a large and complex decision-making problem it is essential to break it down as a hierarchy. The AHP allows that.

The AHP defines a process for decision-making – Formal processes for decision-making are the need of the hour. Decisions, especially collective ones, need to evolve. A process is required that will incorporate the decision-maker's inputs, revisions and learnings and communicate them to others so as to reach a collective decision. The AHP has been created to formalise the process and place it on a scientific footing. The AHP helps in aiding the natural decision-making process.

2.6 The AHP – Applications

Since its discovery the AHP has been applied in a variety of decision-making scenarios:

- Choice – selection of one alternative from a set of alternatives.
- Prioritisation/evaluation – determining the relative merit of a set of alternatives.

- Resource allocation – finding best combination of alternatives subject to a variety of constraints.
- Benchmarking – of processes or systems with other, known processes or systems.
- Quality management.

Domains that have seen many applications of the AHP include healthcare, defence, project planning, technological forecasting, marketing, new product pricing, economic forecasting, policy evaluation, social sciences, etc. Besides its applications in conflict analysis, military operations research, regional and urban planning, R&D management and space exploration, the AHP has developed as a widely accepted methodology for decision-making. As a technique it has evolved over the years and has been applied in conjunction with other mathematical modeling and analysis techniques.

2.7 Pitfalls, Modifications and Extensions

Despite wide applications of the AHP in a variety of domains and at different levels of the decision hierarchy, the AHP has been criticized from several viewpoints. The first problem is that of rank reversal. This was indicated by [12]. In many scenarios, the rankings of alternatives obtained by the AHP may change if a new alternative is added. Belton and Gear introduced one alternative, which was an exact copy of one of the alternatives and then re-evaluated the matrices. This amounted to adding one more column to the matrix with elements similar to those of the original entries in the column corresponding to the earlier alternative.

Robins [2, 3, 4] enumerates the following five issues related to the application of the AHP:

- Vendors get improperly penalized.
- The ratio scale is inaccurate.
- The process can generate inconsistencies as an artefact of its calculations that have nothing to do with consistency of judgment.
- Rank reversal.

The AHP has seen major controversies. One of them has been reflected in the exchanges of Dyer with Saaty and Vargas [8, 9, 10, 11], in the journal *Management Science*.

Despite the controversies and problems faced by the technique of the AHP, it has survived and thrived. Its ease of use and widespread acceptance has resulted in it being applied to decisions related to war games, to technology forecasting, to the evaluation of attack helicopters, to the assessment of presidential candidates, to decisions about buying a car, to choosing one's spouse. In the next chapters we will focus on how the AHP can be used to aid strategic-level decisions in business, defence and governance.

References

1. Triantaphyllou E, et al. (1998) Multi-criteria decision-making: An operations research approach. In: Webster JG (ed.) Encyclopedia of Electrical and Electronics Engineering. Wiley, New York, 15: 175–186.

2. Robins ES, Five Major Pitfalls in the AHP Process, Technical Report no. 9811pub-esr, http://www.TechnologyEvaluation.com
3. Robins ES, The Analytic Hierarchy Process – Issues, Problems, and Recommendations, Technical Report no. 9811pub-esr, http://www.TechnologyEvaluation.com
4. Robins ES, An Investigation into the Efficacy of the Consistency Ratio with Matrix Order – Limits of the AHP, Report no. ARL97-ER-D01, http://www.TechnologyEvaluation.com
5. Triantaphyllou E, Mann SH (1989) An examination of the effectiveness of multi-dimensional decision-making methods: a decision-making paradox. Decision Support Systems 5: 303–312.
6. Forman EH, Gass SI (2001) The analytic hierarchy process – an exposition. Operations Research 49(4): 469–486.
7. Winkler RL (1990) Decision modeling and rational choice: AHP and utility theory. Management Science 36(3).
8. Dyer JS (1990) Remarks on the analytic hierarchy process. Management Science 36(3).
9. Saaty TL (1990) An exposition of the AHP in reply to the paper "Remarks on the analytic hierarchy process". Management Science 36(3).
10. Harkar PT, Vargas LG (1990) Reply to "Remarks on the analytic hierarchy process" by J S Dyer. Management Science 36(3).
11. Dyer JS (1990) A clarification of "Remarks on the analytic hierarchy process", Management Science 36(3).
12. Belton V, Gear AE (1983) On a shortcoming of Saaty's method of analytic hierarchies. Omega 11(3): 227–230.
13. Saaty TL, Vargas LG (1984) The legitimacy of rank reversal. Omega 12(5): 513–516.
14. Belton V, Gear AE (1985) The legitimacy of rank reversal – a comment. Omega 13(3): 143–144.
15. Vargas LG (1985) A rejoinder. Omega 13(4): 249.
16. Zeshui X, Cuiping W (1999) A consistency improving method in the analytic hierarchy process. European Journal of Operational Research 116: 443–449.
17. Ramanathan R (1997) Stochastic decision-making using multiplicative AHP. European Journal of Operational Research 97: 543–549.
18. Frei FX, Harker PT (1999) Measuring aggregate process performance using AHP. European Journal of Operational Research 116: 436–442.
19. Saaty TL, Vargas LG (2001) Models, Methods, Concepts and Applications of the Analytic Hierarchy Process. Kluwer, Dordrecht.
20. Saaty TL (1980) The Analytic Hierarchy Process. McGraw-Hill, New York.
21. Saaty TL, Vargas LG (1991) Prediction, Projection and Forecasting. Kluwer, Boston, MA.
22. Golden BL, Wasil EA, Harker PT (eds) (1989) The Analytic Hierarchy Process. Springer, Berlin.

PART 2

Strategic Decision-Making in Business

3 Aligning Strategic Initiatives with Enterprise Vision

People and their managers are working so hard to be sure things are done right, that they hardly have time to decide if they are doing the right things.
(Stephen Covey)

3.1 Introduction

Doing the right things, the right way, right on target and achieving more with less requires formulating and deploying sound strategies. Today's fierce global competition demands excellence both in strategy and in its execution by senior management in order to meet the challenges of tomorrow. The act of balancing strategy and operations, and continual worry about the future, always push the top management to the helm. Unless there is a common and mutually agreed rational framework that helps align the various units at work in an enterprise with vision from conception till declaration of results, it is not possible to build long-lasting balanced organizations.

Balanced scorecard (BSC), originally developed in the early 1990s by Robert Kaplan and David Norton, is one such framework that helps achieve the required balance. It helps translate the strategy into actions from four perspectives:

- *Financial*: Traditional measures of profitability, revenue, and sales growth.
- *Customer*: Customer retention, customer satisfaction and market research.

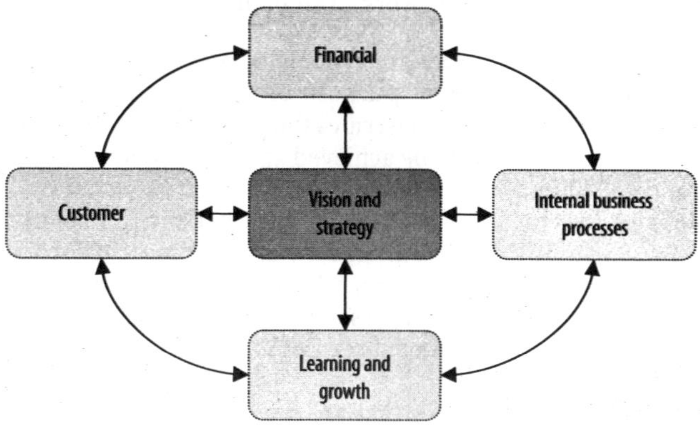

Figure 3.1 Balanced scorecard.

- *Internal business processes*: Processes to meet or exceed customer expectation.
- *Learning and growth*: How the organisation and its people grow and meet new challenges?

BSC is different from the traditional performance management system, because:

1. It also includes non-financial measures for evaluating the overall performance of an organisation.
2. It brings in the concept of defining leading and lagging performance indicators/drivers to compare past and plan future performance targets.
3. It includes indicators from both internal and external stakeholder perspectives and helps build a balance between them.
4. It acts as an effective communication vehicle, translates the strategy into focused measurable actions and aligns the entire organization with the vision.

3.2 A Framework to Align Strategic Initiatives with Vision

Major changes take a long time in big organizations and involve people across the hierarchies. It is thus important to have a framework that fits the following description:

1. It is consistent, robust and stable. It takes from management the burden of ensuring timeliness and consistency in its message. It helps senior management remain focused on the implementation of these change initiatives without getting deflected by the internal or external chaos.
2. It evaluates strategies over its life cycle and realigns them whenever there is a shift.
3. It acts as a common communication vehicle across the enterprise from top to bottom.
4. It takes less effort, and enables quick and accurate delivery of key information while aligning various layers and locations of an organization on an end result.
5. It is flexible enough to operate in a constantly changing environment. As conditions change, it tests current strategies and inducts new ones. It navigates the entire organization during the course of rapid and chaotic changes.

The following steps describe a framework to align initiatives with enterprise vision:

Step 1: Mission, vision and values are articulated and communicated across the organisation to ensure the constant purpose of existence and progress towards excellence in performance. Mission describes the purpose of existence of an organisation, vision defines the results to be achieved and values define the guiding principles or the code of ethics for the organisation.

Step 2: Senior management assesses (a) the external environment (current and future market opportunities, competition etc.) and (b) the internal environment (current strengths and weaknesses), and brainstorms to formulate strategies. Strategy is the approach (based on the assessment) to accomplish the mission and implement the vision. Formulation of strategies is a step forward towards the execution of vision.

Step 3: Define strategic objectives that are measurable. Performance measures are assigned to these strategic objectives. Such measures represent the performance outcome objectively. Targets are set to drive the performance outcome of the organisation. For instance, if the strategy is "Become customer-driven", the strategic objectives could be (a) to assure the timely delivery of the solution and (b) to enhance the quality level of the solution. The performance measures for the objectives could be (a) a percentage of on-time deliveries to a customer and (b) a percentage of defective deliveries. And targets could be (a) 100% in the coming year and (b) zero defects in the year after.

Tying performance measures to the objective is the most critical step towards organisational alignment. It is advisable to involve all stakeholders when the performance measures are being defined. It helps building buy-in and incorporating their feedback at an early stage. However, it is management's responsibility to ensure consistency in the definition and deployment of the measures across the organisation.

There are numerous ways for an organisation to identify the right performance measures for the strategic objectives. These include process modelling, simulation (e.g. systems thinking), a value chain analysis, external benchmarking and cause-effect analysis (e.g. fishbone diagrams). External benchmarking (a) guides management while setting targets based on the peer performance and gap analysis, (b) validates the extent of improvement achievable realistically in a given time frame and (c) provides insights into industry best practices (potential initiatives).

Step 4: The strategic objectives may be mapped to the four perspectives as prescribed by balanced scorecard (BSC) and/or to the total quality management (TQM) themes of overall performance excellence such as a Malcolm Baldrige National Quality Award (MBNQA). The combination of BSC and performance excellence models provides a tremendous advantage to an organisation. Table 3.1 illustrates how the combination helps complement each other.

Mapping to the overall performance excellence themes helps identify gaps and check whether the business strategy is balanced or not. Table 3.2 illustrates how the BSC initiatives find a strategic fit to the performance excellence theme and help the organisation to remain balanced.

Step 5: Initiatives are identified, implemented and managed by metrics and targets to ensure that they are successful. The success of each initiative is measured based on the performance outcome. This performance outcome can be benchmarked against the best-in-class organisations.

Table 3.1 Combining quality and strategy models.

	Balanced scorecard	*Performance excellence model (e.g. MBNQA)*
Purpose and constructs	• Performance measurement and management • Hypothesis-driven	• Total quality management • Fact-based
Key activities	• Links strategy, objectives, initiatives • Sets performance measures and target	• Benchmarking and assessment • Identification of key areas for improvement
Key drivers	• Vision- and strategy-driven • Direction setting	• Excellence in execution driven • Continuous improvement
Key focus	• Governance processes • To-be viewpoint	• Ongoing daily operational process • As-is viewpoint
Commitment level and locus of control	• Senior management	• Senior management

Table 3.2 Complementary models to execute strategies and align the organisation.

Performance excellence themes (e.g. MBNQA)	*Illustrative strategic objectives (balanced scorecard)*
Business results	• Increase business revenues • Increase return on equity • Improve market share of breadwinner products
Process management	• Improve efficiency of operations • Reduce cost of production • Build production capacity • Improve process capability
Customer and market focus	• Extend product portfolio • Enrich existing product portfolio • Expand geographic reach • Improve lifetime profitability per customer
Human resource focus	• Increase employee motivation and satisfaction • Reduce turnover rate • Induct and strengthen strategic skills and competencies
Leadership	• Strengthen corporate governance system and structure • Development of leaders • Building high performance and ethical workplace environment
Measurement, analysis and knowledge management	• Improve security and confidentiality of information • Improve validity, accuracy and timeliness of performance measures
Strategic planning	• Improve competitive intelligence availability and completeness

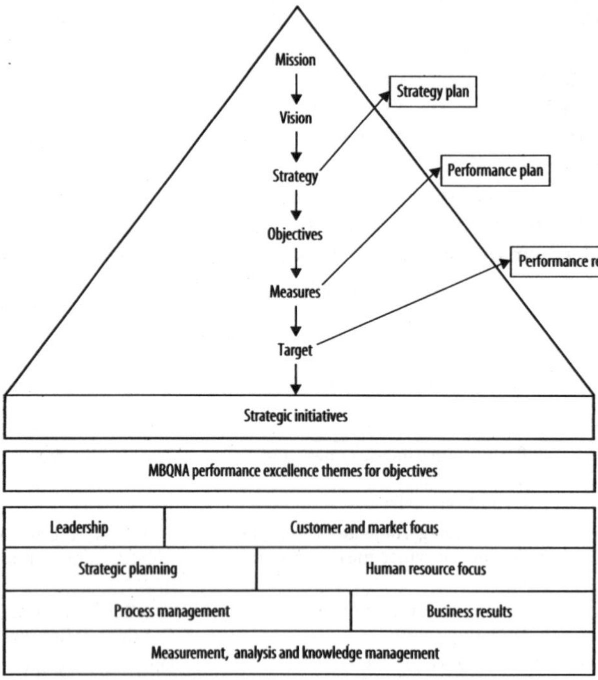

Figure 3.2 Aligning initiatives with enterprise vision: a framework.

Step 6: Initiatives are prioritised, resources are committed and the budget is set to launch and run the initiatives. The initiatives are tracked on a periodic basis to check the progress, the deviations from the budget and the percentage completion rate. The initiatives can be prioritised, suspended or abandoned according to their relative performances any change in strategic focus. The rewards and compensation are tied to the success of these initiatives to motivate the departments, teams and individuals running the initiatives.

Implementing such a framework is itself a big change initiative. Having a framework that helps manage many important initiatives comes with its own overhead to maintain and manage a disciplined approach to decision-making. This constrains the management at times when the tides of change are rapid. Implementation of a framework is ineffective if the management culture is to keep rushing to judgements and bypassing the framework. As a practice, the framework is never rolled out across the organisation until it gets deeply rooted into the management culture and significant signs of continuous maintenance and respect are visible. Only then can strict adherence to the framework by middle and line management be expected.

3.3 Consensus and Prioritisation Issues in Initiative Management

Most organisations face massive resistance in getting the objectives agreed and finalised for inclusion in the scorecard. Even after that, there is a lack of adequate buy-in of the decision, which tends to be considered personal and subjective. There is also a tendency to believe that there are hidden agendas, which are not discussed in the open. Apart from the difficulty in reaching a consensus, some of the other challenges include: prioritising the initiatives, defining performance measures objectively, funding long-term versus short-term initiatives, and procedures for measuring the success of initiatives etc.

The AHP has been applied in complex, real world, multi-criteria decision-making problems to evaluate strategic alternatives. Let us consider four objectives for the year ahead for a global organisation (see Table 3.3). The first step is to evaluate how much these four objectives help to fulfil the vision considering the situational circumstances of the organisation.

Paired comparisons of these objectives are obtained from the top management in terms of relative importance to the vision. The comparisons are shown in Table 3.4.

Each of these objectives has 2–3 initiatives that have been identified by the management. These are shown in Table 3.3. The paired comparisons are shown in Tables 3.5, 3.6, 3.7 and 3.8 respectively, corresponding to objectives O1, O2, O3 and O4 based on their capability of execution to meet the targets.

Final global priorities for these initiatives with respect to the vision come out to be as shown in Table 3.9.

According to the above analysis, the enterprise needs to deploy its major effort in its R&D program (global priority = 0.369), and focus on sales promotion initiatives (global priority = 0.147) and new product development (global priority = 0.141). Hence it is clear that extending the product portfolio along with sales promotion initiatives should be backed up by deep and committed investment in

Table 3.3 Aligning initiatives with enterprise vision.

Theme	*Objectives*	*Measures*	*Target*	*Initiatives*
Financial perspective	• Increase business revenues and profitability (O1)	• Sales growth • Profitability	>30% CAGR >15% CAGR	• Sales promotion (I11) • Enter new geographies (I12) • New marketing channel (I13)
Internal processes	• Improve efficiency of operations (O2)	• Number of defective products	<1% of total	• Six sigma (I21) • Re-engineering (I22)
Customer and market perspective	• Extend product portfolio (O3)	• Number of new product lines	2 new products	• Product development (I31) • R&D program (I32) • Product re-engineering (I33)
Learning and growth	• Increase employee motivation and satisfaction (O4)	• Employee survey • Turnover rate	>99% satisfied <3%	• Restructure compensation (I41) • Cross-train (I42) • Knowledge management (I43) • Personal development opportunities (I44)

Table 3.4 Paired comparisons of objectives.

Objective	*O1*	*O2*	*O3*	*O4*	*NEV*
O1	1	2	1/3	4	0.231
O2	1/2	1	1/5	1/6	0.066
O3	3	5	1	8	0.564
O4	1/4	6	1/8	1	0.139

Table 3.5 Initiatives corresponding to objective O1.

Initiative	*I11*	*I12*	*I13*	*NEV*
I11	1	3	5	0.637
I12	1/3	1	3	0.258
I13	1/5	1/3	1	0.105

Table 3.6 Initiatives corresponding to objective O2.

Initiative	*I21*	*I22*	*NEV*
I21	1	3	0.75
I22	1/3	1	0.25

Table 3.7 Initiatives corresponding to objective O3.

Initiative	*I31*	*I32*	*I33*	*NEV*
I31	1	1/3	3	0.250
I32	3	1	6	0.655
I33	1/3	1/6	1	0.095

Table 3.8 Initiatives corresponding to objective O4.

Initiative	*I41*	*I42*	*I43*	*I44*	*NEV*
I41	1	1/3	3	4	0.243
I42	3	1	6	8	0.596
I43	1/3	1/6	1	2	0.099
I44	1/4	1/8	1/2	1	0.061

the R&D program. These initiatives should be followed up by the cross-training of the workforce and entering new geographies.

The AHP-based framework, as described above, can help prioritize the various enterprise initiatives and also help the alignment of the various initiatives with the vision. If need be, the initiatives can be prioritized on multiple criteria such as competitive advantage, strategic fit, return on investment, payback period, technical risk, business risk, organizational risk etc.

Table 3.9 Final ranking of initiatives.

Objectives	*Initiatives*	*Local rating*	*Global rating*
O1: Increase business revenue; wt: 0.231	I11: Sales promotion	0.637	0.147
	I12: Enter new geographies	0.258	0.060
	I13: New marketing channel	0.105	0.024
O2: Operational efficiency; wt: 0.066	I21: Deploy six sigma	0.75	0.050
	I22: Re-engineering	0.25	0.017
O3: Extend product portfolio; wt: 0.564	I31: Product development	0.250	0.141
	I32: R&D program	0.655	0.369
	I33: Product re-engineering	0.095	0.054
O4: Increase employee motivation and satisfaction; wt: 0.139	I41: Restructure compensation	0.243	0.034
	I42: Cross-train	0.596	0.083
	I43: Knowledge management	0.099	0.014
	I44: Personal development	0.061	0.008

References

1. Kaplan RS, Norton DP Translating Strategy into Action: The Balanced Scorecard. Harvard Business School Press, Boston, MA.
2. Murphy T Achieving Business Value From Technology. Wiley, Hoboken, NJ.
3. Young SD, O'Byrne SF EVA and Value-based Management. McGraw-Hill.
4. National Institute of Standards and Technology. Criteria for Performance Excellence, Baldrige National Quality Program, www.quality.nist.gov.

4 Evaluating Technology Proliferation at Global Level

4.1 Mobile Technologies

The fast-paced future business environments will require applications exploiting technologies that provide network connectivity on the move. The enabling technologies are grouped into mobile computing technologies. A strategic question that needs to be answered for a company investing in mobile technologies, creating a product based on mobile technologies, or providing mobile computing and communication services for the global scenario, is how and in what form these technologies will proliferate round the world. The question assumes greater importance because of the rapidly changing nature of mobile technologies and the variations in investment resulting from them. The nature and proliferation of mobile computing applications varies with geography and culture across the globe. The availability of mobile service providers, good payment models, security, mobile access devices, technology and content will have an impact on mobile computing applications. For a global company it is of paramount importance to evaluate how and in what form a new and emerging technology such as mobile computing applications will proliferate so as to design its services to maximise the revenues and minimise the risk.

In this chapter we describe a three-dimensional matrix approach, based on the analytic hierarchy process (AHP) to evaluate various mobile computing applications. This approach is generic and can be used to evaluate specific applications. The three dimensions of geographic and cultural characteristics, industry segments and population segments will affect the chances of any particular application becoming a killer application.

4.2 Mobile Computing Applications

With enhanced and available communications the motivation to buy and trade while moving will increase. However, initial inertia will hinder people from committing to mobile computing applications in a major way, owing to constraints of security and performance, and also uncertainty. Therefore, initial applications in mobile computing will be simple, location- or time-based service/information provisioning and mobile intranets available to employees on the move. Once customers gain confidence in the technology, they will come forward to make payments and transactions through the mobile internet. There are four categories of applications that have been discussed in the literature, namely transaction management, digital content delivery, telemetry services and passive applications [1].

Transaction management includes online shopping, real-time purchase of or payment for services and micro transactions where e-cash-equipped PDAs and mobile phones will be used to settle transactions such as subway fees. Content delivery services are information browsing, including directory services such as Yellow Pages, and video content for entertainment, product demonstrations and distance learning. Telemetry services include transmission and receipt of sensing, status and measurement information, which will lead to appliance control through mobile handheld devices. Passive applications include recording payment of toll, mass-transit, fast-food and other transactions without user authorisation or entering an identity at each transaction. Some generic applications are described below.

4.2.1 Customer-specific Information Services

The enhanced customer focus will drive the service providers to focus on customer-specific information provisioning related to sports, entertainment, banking, financial markets and location-based services.

4.2.2 Infostations

These stations will contain massive amounts of information and will be available at all possible sites: airports, hotels, railway platforms, consumer stores, cinema halls, etc. As soon as a subscriber to a station passes nearby with his handheld PDA or mobile device, the station will enrich the mobile device with current updates of the relevant information.

4.2.3 In-vehicle Information System (IVIS)

Future vehicles will come tailor-made with IVIS. The IVIS will be a wireless link to a service provider who has a tieup with the vehicle manufacturer or dealer. The IVIS will help in roadside assistance, push location-based information to the vehicle, help in stolen vehicle tracking, provide route assistance, help in vehicle diagnostics and provide web-based services.

A killer application is defined as an application based on a mobile network which will be desired by all users of the network. Three critical dimensions, i.e. geography and culture, segment of population, and industry segment, influence mobile applications.

4.3 The Three Dimensions

Geography and culture is the first dimension that will drive the chances of a mobile computing application becoming a killer application. This is more evident from the study of percentage of internet users accessing specific sites in the USA, Europe, and Asia. The study is based on data available at www.netvalue.com. The study indicates that web users prefer more of local content when accessing the internet through their desktops, and this will be even more the case once mobile users are accessing the wireless internet. The natural delineation reflected in the USA, Europe and Asia–Pacific are considered as the demarcation boundaries of this dimension [2].

The second dimension of the evaluation matrix is the segment of the population which will be accessing the wireless internet. Three population segments, i.e. youth, business people and common people, are considered for this dimension.

The third dimension for evaluating mobile commerce is the industry segment. This segment will embrace mobile technologies as per their own requirements and needs of their customers. For example, travel and transport segments may be the first and most prevalent acceptors of mobile technologies, as this naturally fits into their basic business based on mobility. In this chapter four industry segments are considered, namely travel, finance, entertainment and information, and retail and distribution.

4.4 Killer Sure Scores of Mobile Computing Apps Using the AHP

The methodology now illustrated is for computing KiSS scores for infostations. The hierarchy of comparison for any application is given in Figure 4.1 as the first step in the AHP.

The next step is to evaluate the extent of mobile internet access in various regions of the world. For this, experts are asked to complete the pairwise formats to compute the relative extent of mobile internet penetration in the USA, Europe and Asia–Pacific.

The comparison matrix is shown in Table 4.1. Similarly computed are pairwise comparisons of mobile internet penetration across various regions (Tables 4.2, 4.3 and 4.4) and pairwise comparisons of the need for infostations by different industry segments (Tables 4.5, 4.6 and 4.7). However, in the example it is done only for a single

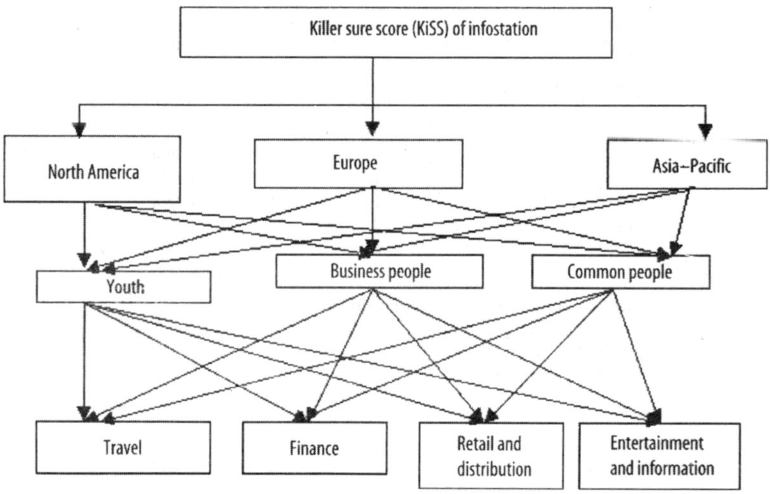

Figure 4.1 AHP hierarchy for evaluating killer sure scores.

Table 4.1 Comparison matrix of geographic areas (mobile internet penetration).

	NA	*E*	*AP*	*NEV*
NA	1	1/3	1/2	0.169
E	3	1	1	0.443
AP	2	1	1	0.387

Table 4.2 Comparison matrix of population segment in North America (mobile internet penetration).

	Y	*BP*	*CP*	*NEV*
Y	1	1/3	2	0.249
BP	3	1	3	0.594
CP	1/2	1/3	1	0.157

Table 4.3 Comparison matrix of population segment in Europe (mobile internet penetration).

	Y	*BP*	*CP*	*NEV*
Y	1	1/5	1/3	0.109
BP	5	1	2	0.582
CP	3	1/2	1	0.309

Table 4.4 Comparison matrix of population segment in Asia–Pacific (mobile internet penetration).

	Y	*BP*	*CP*	*NEV*
Y	1	3	5	0.637
BP	1/3	1	3	0.258
CP	1/5	1/3	1	0.105

Table 4.5 Comparison matrix of infostation access through mobile devices by youth with respect to industry segment.

	T	*F*	*R&D*	*E&I*	*NEV*
T	1	5	3	1/5	0.204
F	1/5	1	1/3	1/9	0.046
R&D	1/3	3	1	1/7	0.094
E&I	5	9	7	1	0.657

Table 4.6 Pairwise comparison of infostation access through mobile devices by business people with respect to industry segment.

	T	*F*	*R&D*	*E&I*	*NEV*
T	1	2	3	7	0.475
F	1/2	1	2	7	0.300
R&D	1/3	1/2	1	5	0.177
E&I	1/7	1/7	1/5	1	0.047

Table 4.7 Pairwise comparison of infostation access through mobile devices by common people with respect to industry segment.

	T	*F*	*R&D*	*E&I*	*NEV*
T	1	7	5	1/3	0.290
F	1/7	1	1/2	1/9	0.047
R&D	1/5	2	1	1/7	0.076
E&I	3	9	7	1	0.587

Table 4.8 Variables to calculate the KiSS score for North America.

	North America (w = 0.169)		
	Y (a_1 = 0.249)	*BP (a_2 = 0.594)*	*CP (a_3 = 0.157)*
T	$X_{11} = 0.204$	$X_{12} = 0.475$	$X_{13} = 0.290$
F	$X_{21} = 0.046$	$X_{22} = 0.300$	$X_{23} = 0.047$
R&D	$X_{31} = 0.094$	$X_{32} = 0.177$	$X_{33} = 0.076$
E&I	$X_{41} = 0.657$	$X_{42} = 0.047$	$X_{43} = 0.587$

Table 4.9 Killer sure scores of infostation in various business sectors in North America.

	KiSS scores for infostation in North America = ($w \times a_j \times X_{ij}$) × 1000		
	Youth	*Business people*	*Common people*
Travel	8.58	47.68	7.69
Finance	1.94	30.12	1.25
R&D	3.96	17.77	2.02
E&I	27.65	4.72	15.57

region, and the results assumed applicable to all. (Abbreviations used in the tables are: NA, North America; E, Europe; AP, Asia–Pacific; Y, youth; BP, business people; CP, common people; T, travel industry; F, finance industry; R&D, retail and distribution; E&I, entertainment and information industry.)

The next step is to calculate the normalized eigenvector, which gives the relative ranking of these regions. Consistencies of these matrices were checked and CR was found to be less than 0.1.

These rankings are combined to get the killer sure score of infostation in various business sectors. Table 4.8 gives the variables required for computing the KiSS score for North America and Table 4.9 the KiSS scores for North America.

Table 4.10 shows the relative embracing of mobile technologies in North America, Europe and Asia-Pacific according to population segment and industry segment.

Kiviat charts corresponding to North America, Europe and Asia–Pacific are shown in Figures 4.2, 4.3 and 4.4 respectively. We can see from these that it is in North America that the likelihood of business people accessing travel-related information (BPT) on their mobile devices is largest. Business people accessing finance information comes second and youth accessing infotainment services comes close third. In Europe too, business people accessing travel-related information comes first.

Table 4.10 Relative embracing of mobile technologies in North America, Europe and Asia–Pacific with respect to population segment and industry segment.

Population and industry segment	*North America*	*Europe*	*Asia–Pacific*
Youth accessing travel information (YT)	0.051	0.022	0.130
Youth accessing financial information (YF)	0.011	0.005	0.029
Youth accessing retail and distribution information (YR&D)	0.023	0.010	0.060
Youth accessing entertainment and information (YE&I)	0.164	0.072	0.419
Business people accessing travel information (BPT)	0.282	0.276	0.123
Business people accessing financial information (BPF)	0.178	0.175	0.077
Business people accessing retail and distribution information (BPR&D)	0.105	0.103	0.046
Business people accessing entertainment and information (BPE&I)	0.028	0.027	0.012
Common people accessing travel information (CPT)	0.046	0.090	0.030
Common people accessing financial information (CPF)	0.007	0.015	0.005
Common people accessing retail and distribution information (CPR&D)	0.012	0.023	0.008
Common people accessing entertainment and information (CPE&I)	0.092	0.181	0.062

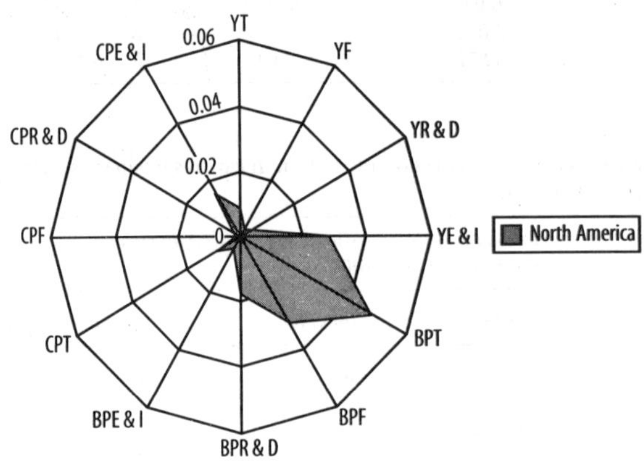

Figure 4.2 Mobile computing Kiviat chart for North America.

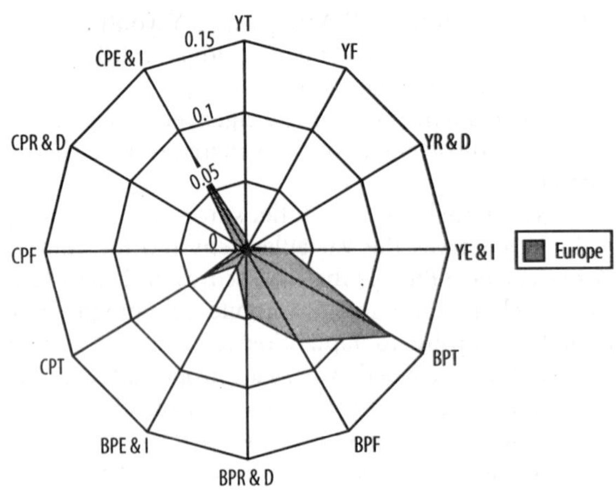

Figure 4.3 Mobile computing Kiviat chart for Europe.

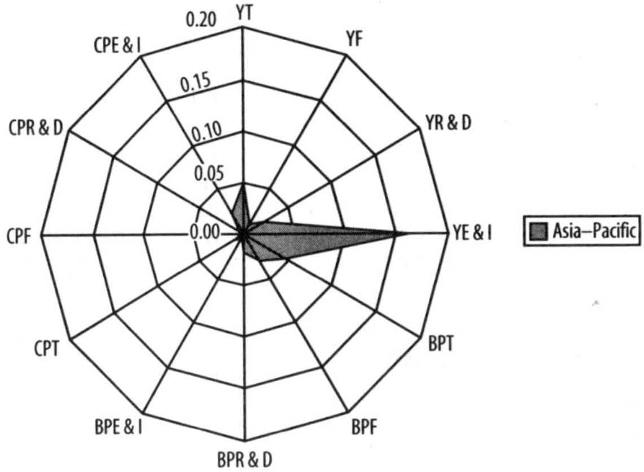

Figure 4.4 Mobile computing Kiviat chart for Asia–Pacific.

Table 4.11 KiSS categories.

Category of application	*KiSS score*
Frozen (F)	<5.0
Cold (C)	≥5.0; <20.0
Warm (W)	≥20.0 <50.0
Hot (H)	≥50.0; <100.0
Red-hot (R)	≥100.0

Table 4.12 Region, population segment and industry segment distribution of KiSS for infostation application.

	North America			*Europe*			*Asia–Pacific*		
	Y	*BP*	*CP*	*Y*	*BP*	*CP*	*Y*	*BP*	*CP*
T	C	W	C	C	R	W	W	W	C
F	F	W	F	F	H	C	C	W	F
R&D	F	C	F	F	W	C	W	C	F
E&I	W	F	C	W	C	H	R	F	W

However, at the number two spot comes common people accessing infotainment services, followed by business people accessing finance information on their mobile devices. In Asia–Pacific mobile applications are driven by youth accessing infotainment services and travel-related services. Business people accessing travel-related services on their mobile devices comes third.

The next step is to associate the killer sure score with the categories that can be used for drawing the inference. Table 4.11 gives the KiSS according to information category. Five categories of KiSS are defined for application evaluation: frozen, cold, warm, hot and red-hot. Table 4.12 gives the results of the analysis for all the regions.

Table 4.13 KiSS analysis of applications for various verticals in Europe.

	Travel			*Finance*			*E&I*		
	Y	*BP*	*CP*	*Y*	*BP*	*CP*	*Y*	*BP*	*CP*
Infostation	C	R	W	F	H	C	W	C	C
IVIS	R	R	H	F	H	F	R	C	W
Voice portal	R	H	R	C	R	W	R	W	H
Profile matching	H	F	C	F	H	C	R	C	H

4.6 Strategic Decision Guidance

Let us assume that a mobile service provider (MSP) wants to deploy mobile services in, say, Europe. The provider is presently thinking of, say, the following applications: infostations, in-vehicle information services (IVIS), voice portal and profile matching. For the sake of simplicity let us assume that the KiSS analysis of these applications for various verticals in Europe has yielded the results shown in Table 4.13.

Further, let us assume that the MSP subscribers are in the following ratio: youth 25%, business people 35% and common people 40%. However the revenues generated from the three classes are in the ratio 20%, 50% and 30% respectively. The cumulative normalised weight of youth, business people and common people comes out to at 14%, 51% and 35% respectively. The question the MSP needs to answer is what content they should provide in relation to the applications planned. Also, what should be the order of deployment of these applications? Using the above analysis and the cumulative weights we can see that the most profitable deployment for the MSP would be IVIS, with travel-related services for every type of customer. Second in line is a voice portal for every type of customer. Infostations with travel-related services for business people comes next in line.

The above analysis shows how various decisions regarding applications deployment can be taken by the MSP.

Mobile computing killer applications will vary from region to region, from people to people and from industry segment to industry segment. Application that is successful in one region may not be a success in another. Scientific analysis based on the three-dimensional matrix approach and the AHP, as described in this chapter, will provide the analysis for the mobile service provider to develop and deploy applications that can become killer in their area of operation.

References

1. Senn JA. (2000) The emergence of m-commerce. IEEE Computer, December.
2. Agoston TC, Uedat, Nishimura. (2000) Pervasive computing in a networked world. Proceedings for INET. (http://www.isoc.org)

5 Evaluating Enterprise-wide Wireless Adoption Strategies

5.1 Mobile Technologies

The promise of mobile technologies and in turn mobile business and commerce has yet to be realised; the early hype is now filtering through the sieve of reality. However, while converting rapidly changing mobile technologies into a profitable opportunity is proving to be a walk through quicksand, nevertheless enterprises can ill afford to let mobile technology go. Many analysts, technocrats and business gurus have come up with methods to embrace the mobile channel, indicating that a perfect solution to every possible situation is impossible. We propose an AHP-based framework to study and choose a relevant wireless adoption strategy for the large, global and multilocation enterprises.

This framework allows the strategists and the CEOs of large enterprises to go ahead systematically in evaluating options, strategies and the relevant metrics in order to come out with a comprehensive blueprint for mobile-enabling their enterprise.

The coming together of internet and wireless communications to provide data services for those who are mobile is opening up new ways for enterprises not only to offer their services to their customers, but also to increase the efficiency of their employees. Despite the technology being only nascent, the likely impact of this convergence on industry was foreseen as tremendous. Events and trends however have now reduced this prospect. Expectations remain high, though, if we look at the innovative and focused ways some of the enterprises in Europe have made wireless profitable. This coupled with mass consumer adoption in Japan and other Asian countries provides a ray of hope to service providers, mobile technology investors and mobile product vendors.

In general, mobile access to the internet is increasing rapidly. Enterprises have been making plans to invest in mobile technologies for the mobile enabling of their employees and customers for business purposes. Rather than waiting for the conceptualising of a killer application, it would be better to adapt to wireless within the constraints of the technology. Besides gaining a first-mover advantage, this would offer experience with a technology that is continuously happening. It would also be improving – as new, evolving networks come along based on increasingly powerful industry standards of GPRS, EDGE and UMTS. These are leading to the third generation wireless networks (3G) which can provide up to 2 megabits per second bandwidth. Instead of waiting for technologies to stabilize and then giving mobile delight to customers and employees, it makes sense to deploy mobile applications now.

5.2 Enterprise Wireless Strategies

We propose a step-by-step framework to produce a comprehensive process for mobile adoption in the enterprise.

5.2.1 Identify the Type of Organization

The first step is to identify the type of enterprise. We consider two distinct classes of enterprise – those that have negligible direct consumers (N) and those that are consumer-facing (C). The first class does not have any direct consumers. It includes organizations such as large power generation plants, chemical plants, manufacturing units, etc. In the second class, we find large manufacturers and suppliers of food and beverages, newspaper and magazine publishers, banks, retail stores, hospitals, hotels, airports, etc. It can be seen that consumer-facing enterprises have one major non-controllable parameter, i.e. the mobile devices that their consumers will use. The first class of enterprises can control the mobile devices to their employees and organize their wireless infrastructure to support the specified devices only. In consumer-facing enterprises customer devices constitute an uncertain and sometimes unknown factor.

5.2.2 Evaluate High-priority Targets for Wireless Adoption

Any large enterprise needs to evaluate what are the primary targets for the wireless. Are the employees of the enterprise the main target for mobile adoption or is the target the consumers/customers of the enterprise? In many cases, the tendency is to initiate projects for targets of both types simultaneously. We believe that unless there are significant benefits, it makes sense to start with either consumers or employees as targets. However, in the long run both employees and customers have to be supported. Hence there are three possibilities – only one has to be chosen. These are consumers (C), employees (E) and both (B).

5.2.3 Evaluate and Choose Wireless Strategies

The enterprise has to choose one of the following generic strategies or a mix of them, depending upon the type of organization and type of mobile embracer targeted. Our various strategy categories are given below. Strategies S1 to S3 are related to creating a channel to the customers of the enterprise, while strategies S4 to S6 are related to improving the productivity of employees and the efficiency of the process.

Channel creation (S1): This means creating a new channel across all consumer-facing activities, enhancing the existing business rather than creating new revenue streams.

Revenue transformation (S2): The mobile services may be giving new or more valuable services to the customers above and beyond the core products. This leads to a strategy of creating new revenue streams that complement or even eat up the existing ones.

Piecemeal solution (S3): A piecemeal solution strategy uses the mobile technology as a plug-in, providing solutions to the customers in various pockets, rather than a single, fully integrated solution.

Improving teleworker ratio (S4): A teleworker is an employee who works from home for some fraction of hours in a week using network connectivity with the enterprise's systems. Mobile adoption may mean increasing the ratio of teleworkers in the enterprise. More teleworkers implies lower infrastructure costs, more flexibility for the employers and a higher probability of retaining employees.

Empowering mobile employees (S5): Some employees may need mobile connectivity as a means to access enterprise information systems. Empowering mobile employees means that they will have the capability to access enterprise information, relevant for their immediate task, anywhere, at any time and on the device that they want.

True wireless office – WLAN and wireless WAN (S6): This is quite definitely the ultimate goal for all enterprises, when wireless technologies will connect your enterprise to you and your customers. However, in achieving it each enterprise has to follow the path laid down by the objectives, requirements and resources that it has.

Table 5.1 shows the expected probability of a specific enterprise choosing a particular mobile strategy for specific wireless targets. It can be seen that consumer-facing enterprises are more likely to start with consumer-oriented strategies and then graduate to the final goal of an integrated mobile enterprise. The path to the final goal will depend upon the enterprise's current priorities and the final objective.

Once the main strategy or a mix of strategies, depending upon the requirements, is identified, it has to be sufficiently articulated in terms of the mission, the organization of the strategy implementation team, the skill sets needed or to be acquired, the metrics for measuring the success or otherwise of the strategy and how the risks in implementing the strategy will be taken care of.

Table 5.1 Probability of choosing a particular mobile strategy for the specific organisation and type of wireless target.

Strategy	*Type of organisation and type of wireless target*					
	CC	*CE*	*CB*	*NC*	*NE*	*NB*
S1	High	High	High	Low	High	High
S2	High	Low	High	Low	Low	Low
S3	High	High	High	Low	High	Medium
S4	Low	High	High	Medium	High	High
S5	Low	Medium	Medium	Medium	High	High
S6	Low	Medium	Low	Medium	High	High

CC, consumer-facing organisation targeting consumers; CE, consumer-facing organisation targeting employees; CB, consumer-facing organisation targeting both consumers and employees; NC, no-direct-consumers organisation targeting consumers; NE, no-direct-consumers organisation targeting employees; NB, no-direct-consumers organisation targeting both consumers and employees.

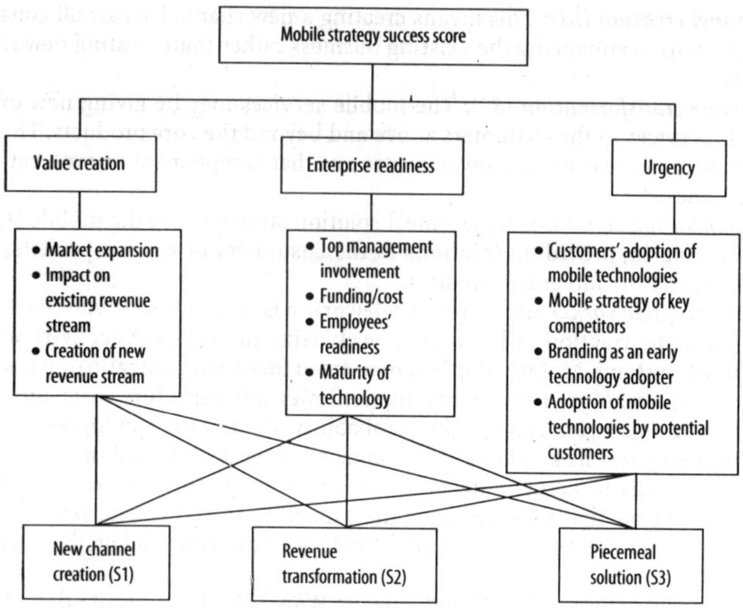

Figure 5.1 Hierarchy for evaluation of customer-facing wireless strategies.

5.3 Factors Affecting Enterprise Wireless Strategies

For evaluating the wireless strategies, three critical parameters need to be looked into. These are value creation, enterprise readiness, and urgency. If the target class for wireless adoption is the customers of the enterprise then the value creation parameter should include the expected market expansion due to the adoption of a particular strategy, the impact on existing revenue streams and the potential for the creation of new revenue streams. The enterprise readiness parameter includes the top management's involvement and readiness, the funding or cost of implementing the strategy, the employees' readiness to absorb the new technology and the maturity of technology for executing the strategy. The urgency parameter includes to what extent customers are adopting mobile technologies, what the current mobile strategy of key competitors is, whether the adoption of a particular strategy will create an image for the enterprise of being an early technology adopter, and what the trends are in the adoption of mobile technologies by potential customers. The hierarchy for the evaluation of customer-facing wireless adoption strategies is shown in Figure 5.1.

If the target class for wireless adoption is the employees of the enterprise then the value creation parameter should include the expected reduction in execution time or increase in throughput of employees due to the adoption of a particular strategy, the expected reduction in cost of infrastructure not related to mobile adoption strategy and the expected improvement in supply chains and/or business processes. The enterprise readiness parameter includes the top management's involvement and readiness, the funding or cost of implementing the strategy, the employees' readiness to absorb the new technology and the maturity of technology

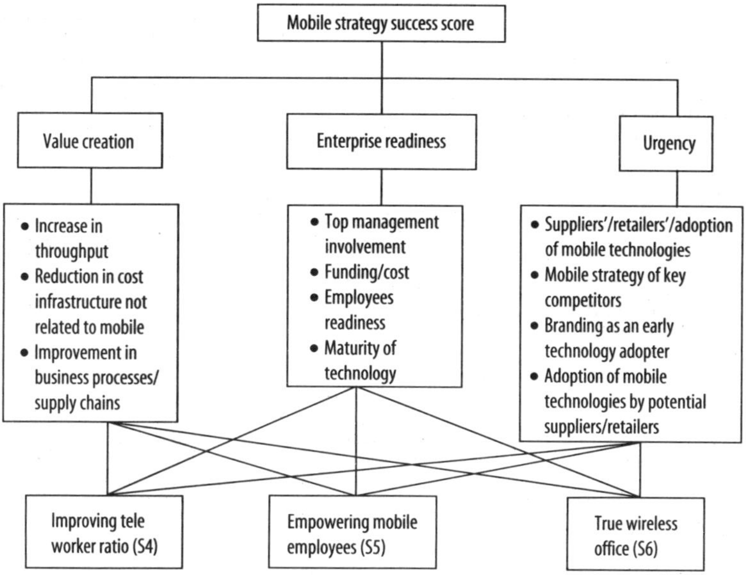

Figure 5.2 Hierarchy for evaluation of employee facing wireless strategies.

for executing the strategy. The urgency parameter includes to what extent the suppliers and retailers are adopting mobile technologies, what the current mobile strategy of key competitors is, whether the adoption of a particular strategy will create an early technology adopter image for the enterprise, and what the trends are in the adoption of mobile technologies by potential suppliers and retailers. The hierarchy for the evaluation of employee-facing wireless adoption strategies is shown in Figure 5.2. As can be seen, the second-level parameters continue to be value creation, enterprise readiness and urgency.

5.4 Using the AHP for Enterprise Wireless Strategies Evaluation

Let us consider a multilocation enterprise whose customers are rapidly adopting mobile technologies. The competitors have already started offering a mobile channel to customers. The suppliers and retailers are also starting to adopt mobile technologies. The employees however display some inertia in adapting to mobile technologies, as they argue that business is not impacted currently and they are able to do their jobs with the existing technologies and processes. The top management does not have much time to devote to the management and implementation of any mobile strategy. However, sufficient funding is available for implementing Phase I of all the six strategies. Now the problem is to prioritize and execute the strategy.

Let the first-level pairwise comparison matrices obtained from top management and other stakeholders of the enterprise for comparing value creation, enterprise readiness and urgency for customer-facing and employee-facing mobile strategies be as given in Tables 5.2 and 5.3 respectively.

Table 5.2 First-level comparison for customer-facing strategies.

	VC	*ER*	*U*	*NEV*
Value creation (VC)	1	1	3	0.429
Enterprise readiness (ER)	1	1	3	0.429
Urgency (U)	1/3	1/3	1	0.143

Table 5.3 First-level comparison for employee-facing strategies.

	VC	*ER*	*U*	*NEV*
Value creation (VC)	1	1/3	3	0.258
Enterprise readiness (ER)	3	1	5	0.637
Urgency (U)	1/3	1/5	1	0.105

Table 5.4 Second-level comparison for the value creation parameter.

	ME	*ERS*	*NRS*	*NEV*
Market expansion (ME)	1	2	3	0.517
Impact on existing revenue stream (ERS)	1/2	1	4	0.359
New revenue stream (NRS)	1/3	1/4	1	0.124

Table 5.5 Second-level comparison for the enterprise readiness parameter.

	TMI	*C*	*ER*	*TM*	*NEV*
Top management involvement (TMI)	1	1/5	3	1/3	0.117
Funding/cost (C)	5	1	7	3	0.565
Employees readiness (ER)	1/3	1/7	1	1/5	0.055
Technology maturity (TM)	3	1/3	5	1	0.262

Table 5.6 Second-level comparison for the urgency parameter.

	CA	*KC*	*BETA*	*PC*	*NEV*
Customers' adoption of mobile technologies (CA)	1	2	2	5	0.450
Mobile strategy of key competitors (KC)	1/2	1	2	3	0.281
Branding as an early technology adopter (BETA)	1/2	1/2	1	2	0.179
Adoption by potential customers (PC)	1/5	1/3	1/2	1	0.090

Let us consider the case of customer-facing wireless strategies first. The second-level parameters for value creation, enterprise readiness and urgency are compared as per Figure 5.1, and are shown in Tables 5.4, 5.5 and 5.6 respectively.

The next step is to pairwise compare the three strategies, i.e. S1, S2 and S3 with respect to each parameter contributing to value creation, enterprise readiness and urgency. These comparisons are shown in Tables 5.7 to 5.17.

The computations required to get the final relative ranking of the three strategies are shown in Table 5.18. The various values used are taken from relative weights and rankings obtained in the above tables.

Table 5.7 Comparison of strategies with respect to market expansion.

	S1	*S2*	*S3*	*NEV*
Channel creation (S1)	1	1/5	3	0.188
Revenue transformation (S2)	5	1	7	0.731
Piecemeal solution (S3)	1/3	1/7	1	0.081

Table 5.8 Comparison of strategies with respect to the existing revenue stream.

	S1	*S2*	*S3*	*NEV*
Channel creation (S1)	1	3	5	0.637
Revenue transformation (S2)	1/3	1	3	0.258
Piecemeal solution (S3)	1/5	1/3	1	0.105

Table 5.9 Comparison of strategies with respect to the new revenue stream.

	S1	*S2*	*S3*	*NEV*
Channel creation (S1)	1	1/5	3	0.183
Revenue transformation (S2)	5	1	8	0.742
Piecemeal solution (S3)	1/3	1/8	1	0.075

Table 5.10 Comparison of strategies with respect to top management's involvement.

	S1	*S2*	*S3*	*NEV*
Channel creation (S1)	1	3	1/5	0.188
Revenue transformation (S2)	3	1	1/7	0.081
Piecemeal solution (S3)	5	7	1	0.731

Table 5.11 Comparison of strategies with respect to funding/cost.

	S1	*S2*	*S3*	*NEV*
Channel creation (S1)	1	5	1/5	0.212
Revenue transformation (S2)	1/5	1	1/8	0.062
Piecemeal solution (S3)	5	8	1	0.726

Table 5.12 Comparison of strategies with respect to employees' readiness.

	S1	*S2*	*S3*	*NEV*
Channel creation (S1)	1	5	1/3	0.279
Revenue transformation (S2)	1/5	1	1/7	0.072
Piecemeal solution (S3)	3	7	1	0.649

Table 5.13 Comparison of strategies with respect to technology maturity.

	S1	*S2*	*S3*	*NEV*
Channel creation (S1)	1	2	1/7	0.131
Revenue transformation (S2)	1/2	1	1/9	0.076
Piecemeal solution (S3)	7	9	1	0.793

Table 5.14 Comparison of strategies with respect to customers' adoption.

	S1	*S2*	*S3*	*NEV*
Channel creation (S1)	1	7	5	0.740
Revenue transformation (S2)	1/7	1	1/2	0.094
Piecemeal solution (S3)	1/5	2	1	0.167

Table 5.15 Comparison of strategies with respect to strategy of key competitors.

	S1	*S2*	*S3*	*NEV*
Channel creation (S1)	1	1/3	3	0.258
Revenue transformation (S2)	3	1	5	0.637
Piecemeal solution (S3)	1/3	1/5	1	0.105

Table 5.16 Comparison of strategies with respect to branding.

	S1	*S2*	*S3*	*NEV*
Channel creation (S1)	1	1	5	0.455
Revenue transformation (S2)	1	1	5	0.455
Piecemeal solution (S3)	1/5	1/5	1	0.090

Table 5.17 Comparison of strategies with respect to adoption by potential customers.

	S1	*S2*	*S3*	*NEV*
Channel creation (S1)	1	1/3	5	0.272
Revenue transformation (S2)	3	1	8	0.661
Piecemeal solution (S3)	1/5	1/8	1	0.067

Table 5.19 shows the relative rankings of each of the strategies with respect to each parameter.

Figure 5.3 shows the Kiviat chart for customer-facing enterprise wireless strategies. As is evident, piecemeal solutions (S3) as per the specific requirements seem to be the best strategy in the given scenario.

A similar analysis can be carried out for wireless strategies which are employee-facing, i.e. for strategies S4 to S6. The final step is to combine both the sets and choose where the enterprise priority lies. For example, let the relative rankings of S4, S5 and S6 come out as 0.45, 0.35 and 0.20 respectively. As we have seen, relative

Table 5.18 Final ratings of customer-facing wireless strategies.

Criterion	*Sub-criterion*	*Weight*	*S1*	*S2*	*S3*
Value creation (VC)	**Relative weight of value creation VC = 0.429**				
	Market expansion (ME)	0.517	0.188	0.731	
	Impact on existing revenue stream (ERS)	0.359	0.637	0.258	0.105
	New revenue stream (NRS)	0.124	0.183	0.742	0.075
	Sub-rating with respect to value creation VC		**0.349**	**0.563**	**0.089**
Enterprise readiness (ER)	**Relative weight of enterprise readiness ER = 0.429**				
	Top management involvement (TMI)	0.117	0.188	0.081	0.731
	Funding/cost (C)	0.565	0.212	0.062	0.726
	Employees' readiness (ER)	0.055	0.279	0.072	0.649
	Technology maturity (TM)	0.262	0.131	0.076	0.793
	Sub-rating with respect to enterprise readiness ER		**0.191**	**0.068**	**0.739**
Urgency (U)	**Relative weight of urgency U = 0.143**				
	Customers' adoption of mobile technologies (CA)	0.450	0.740	0.094	0.167
	Mobile strategy of key competitors (KC)	0.281	0.258	0.637	0.105
	Branding as an early technology adopter (BETA)	0.179	0.455	0.455	0.091
	Adoption by potential customers (PC)	0.090	0.272	0.661	0.067
	Sub-rating with respect to urgency		**0.511**	**0.362**	**0.127**
Final rating			**0.305**	**0.322**	**0.373**

Table 5.19 Relative ranking of various strategies with respect to individual parameters.

Parameter	*Channel creation S1*	*Revenue transformation S2*	*Piecemeal solution S3*
ME	0.0417	0.1621	0.0180
ERS	0.0981	0.0397	0.0162
NRS	0.0097	0.0395	0.0040
TMI	0.0094	0.0041	0.0367
Cost (C)	0.0514	0.0150	0.1760
ER	0.0066	0.0017	0.0153
TM	0.0147	0.0085	0.0891
CA	0.0476	0.0060	0.0107
KC	0.0104	0.0256	0.0042
BETA	0.0116	0.0116	0.0023
PC	0.0035	0.0085	0.0009

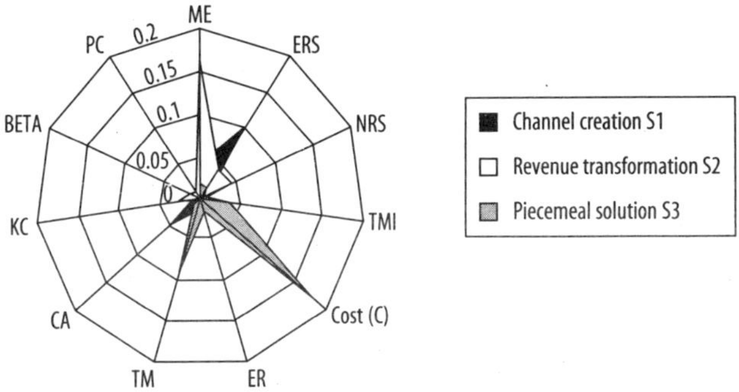

Figure 5.3 Customer-facing wireless strategies.

Table 5.20 Final ranking of enterprise wireless strategies.

Organisation			*Individual rankings*	*Final rankings*
Customer-facing	0.65			
	S1	Channel creation	0.305	0.198
	S2	Revenue transformation	0.322	0.209
	S3	Piecemeal solutions	0.373	0.242
Employee-facing	0.35			
	S4	Improving teleworkers ratio	0.450	0.158
	S5	Empowering mobile workers	0.350	0.123
	S6	True wireless office	0.200	0.070

ranking of customer-facing wireless strategies, i.e. S1, S2 and S3, are 0.305, 0.322 and 0.373 respectively. The next step is to get the relative importance of customer-facing and employee-facing wireless strategies. Let us assume that right now the enterprise top management wants to implement the customer-facing wireless strategy more, with a relative importance of, say, 0.65. The relative importance of the employee-facing wireless strategy then comes out to be 0.35. The final ranking of the six strategies can be computed by multiplying these weights with the ranking of the strategies. For example, for customer-facing strategy S1 the final rank comes out to as $0.65 \times 0.305 = 0.198$. Table 5.20 gives the final ranking of the six strategies.

From the above analysis, it can be seen that if there are sufficient funds then the enterprise can start two initiatives, i.e. providing piecemeal solutions for the customer-facing issues and improving the teleworker ratio. The top management has to take a call and implement an integrated solution based on the above analysis.

6 Software Vendor Evaluation and Package Selection

6.1 Introduction

There are innumerable cases when an organisation has to either abandon the roll-out of a package software solution or else release it midway during the course of customisation. The reasons for such actions could vary: the software vendor closing down its operations, the vendor being acquired (with the consequent uncertainty over the continuation of products and services), a large functional mismatch or feature gaps, very high support and administration costs, etc. An organisation that doesn't adopt a rigorous approach to the evaluation of vendors and selection of a package solution is always open to such risks.

The organisation should be concerned about the vendor and the software evaluation practice. The organisation should invest in adopting best practices to avoid exposure to such risks in future. There are various reasons for this investment:

- Large number of software vendors and packages to choose from.
- Complex or multiple technologies and architectures.
- High economic impact (i.e. cost of acquisition, implementation, maintenance etc.) of decision.
- Technology and business trends are difficult to predict.
- Large and cross-functional teams involved in the buying decision.
- Traditional evaluation methods based on few functional and cost factors, are no longer adequate for complex and multicriteria decision-making.

AHP organises the evaluation criteria in hierarchical order and facilitates the selection of the most suitable alternative. It uses a sound mathematical basis to identify inconsistencies in evaluating scores that may arise due to personal biases, irrational reasoning and deliberate political agenda.

An AHP-based three-step process (see Figure 6.1) of evaluating the software vendor and packaged solution is described in Section 6.2. The activities performed in each of these steps are summarised as follows:

Step 1 – Identifying business requirements: The first step is to identify the business objectives and justify the need of a software-based solution to meet them. Some of the activities in this step are:

- Define the business objectives and requirements.
- Identify potential software vendors and solutions.

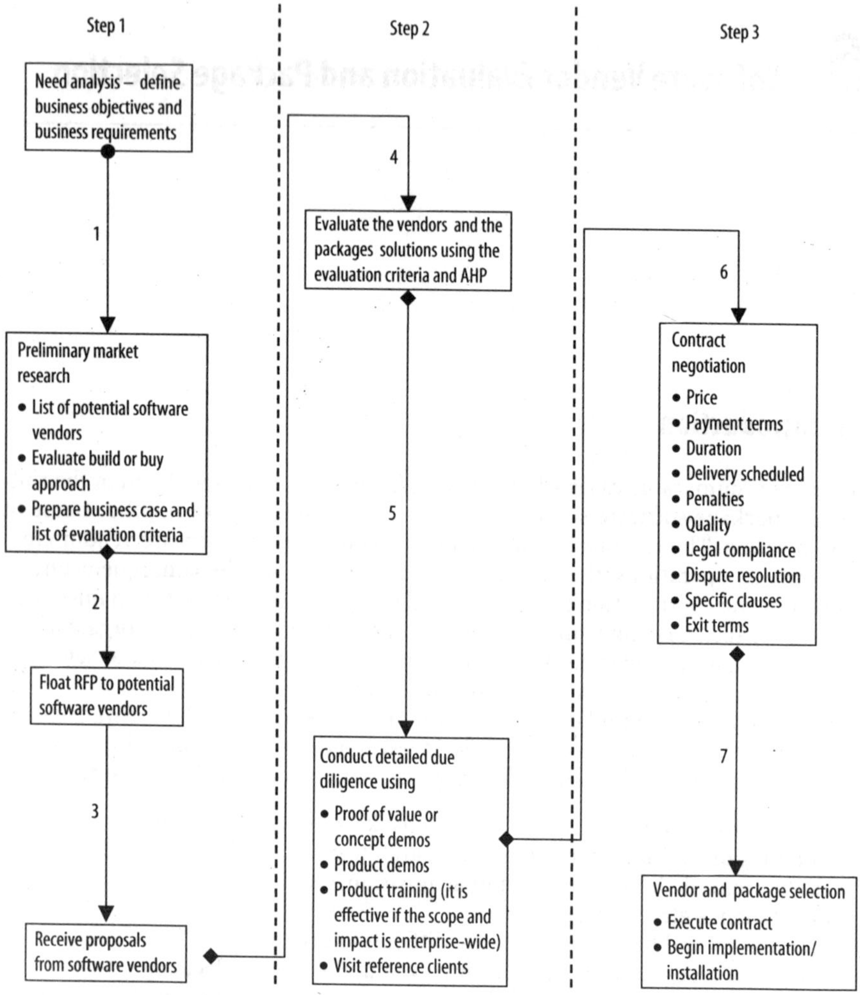

Figure 6.1 Vendor evaluation and selection process.

- Analyse in parallel the alternative approaches to business objectives, i.e. build versus buy.
- Document and send the business requirements as a request for proposal (RFP) to each potential software vendors and wait for proposals.

Step 2 – Evaluate the software vendors and the packaged solutions: The proposed solutions and vendors are evaluated using AHP. The following activities are performed in this step:

- Define the criteria of evaluation for the software vendors and the packaged solutions.
- Apply AHP to evaluate and rank the proposals.
- Conduct detailed due diligence to validate each vendor's credentials and the features of the solutions.

Stage 3 – Recommend: The focus in this step is to optimise the benefits by meticulously negotiating contracts and strengthening the relationship with the software vendor.

Section 6.2 below describes those activities in detail.

6.2 Identifying Business Requirements (Step 1)

The complete and precise understanding of the business objectives and business needs that drive the procurement of the software solution is the key to success. It is essential for the business users, who are going to use the software solution, to participate in elaborating the business requirements with the procurement team. The activity of defining the business objectives and requirements is explained below:

Example

A global company spends huge amounts in IT initiatives to retain its leadership position in the retail market. The IT spend has been exceeding 10% of the total revenues repeatedly for last three years. This percentage is very high when compared with the industry benchmark. The CIO is frequently asked in board meetings to either justify or optimise the IT spend. Economic pressures have forced the company to retain its operating margins to survive the trend of economic downturn. This in turn has put pressure on the CIO to control the IT spend, which has been rising unjustifiably.

The CIO has identified some of the key issues with the current IT organisation, as follows:

- The current IT organisation is decentralised and executes its decisions discretely.
- The IT management team don't maintain a unified repository of assets deployed globally to plan, prioritise and track the spend and its benefits to the business.
- There is no framework to measure and communicate the business value of IT to the organisation.

The CIO has decided to implement a software solution that helps the IT team to overcome such issues. The objective of the solution is to help the IT team in (1) planning, prioritising and tracking the IT initiatives centrally, (2) optimising the spend on IT assets and squeezing out the extra costs from the current IT portfolio, and (3) justifying the business value of IT investments/assets.

Once the business issues and the objectives are known, the next step is to elaborate the business requirements by sitting and discussing with the business users, who will benefit from the software solution. In this case, the senior IT management and the CIO are the business users of the software solution. Based on the interview with them, the requirements are documented and categorised by the evaluation criterion:

A: Business Functionality

Example

A1: The solution should efficiently and effectively support the process of (a) scorecarding the performance, (b) defining IT initiatives, (c) prioritising, executing and tracking the initiatives, and (d) handling exceptions normally encountered during the execution of initiatives, like overruns in budget and implementation time and managing such risks.
A2: The solution should be able to provide a centralised view of IT assets deployed globally.
A3: The IT assets that need to be included for monitoring and tracking the health are (a) IT processes, (b) people, (c) applications and (d) infrastructures.
A4: The software solution must support best practices in space of (a) vendor relationship management, (b) contract negotiation, (c) license management and (d) resource management.

B: Technology

B1 Usability

B1.1 Personalisation

Example

B1.1.1: The IT managers should be able to personalise the screens and organise the work as per their roles.
B1.1.2: The system should support multiple user profiles per user.
B1.1.3: The system should support the national language of the country where it is installed.

B1.2 Accessibility

Example

B1.2.1: The system should be accessible through the existing intranet portal in the enterprise.
B1.2.2: The mobile users, with the system installed on their laptops, should be able to work remotely as well as in offline mode.

B1.3 Learnability

Example

B1.3.1: The design of the user interfaces should be consistent with the existing organisational policy for them.The policy defines the guidelines related to the features such as currency, fonts, reports etc.

B1.3.2: The application should provide context sensitive help to its users.
B1.3.3: Documentation and tutorials.

B2 Architecture

B2.1 Architectural Compliance

Example

B2.1.1: The solution should meet the technology standards as set by the organisation (for network, hardware, OS, database, application server, interfaces and web server etc.).
B2.1.2: Web-based multi-tiered architecture.

B2.2 Scalability

Example

B2.2.1: The system should be scalable to support the current as well as the expected number of users for the next three years.
B2.2.2: The system should be scalable without any performance degradation.

B2.3 Interoperability

Example

B2.3.1: Application to support the existing infrastructure portfolio in the organisation.
B2.3.2: No/minimum variation in performance if the underlying hardware and operating system is changed in future.

B2.4 Performance

Example

B2.4.1: Multiple sessions per user as well as multi-tasking within a session.
B2.4.2: Administration functions for monitoring the performance and the tools to fine-tune it.

B3 Integration

Example

B3.1: Seamless interface with existing groupware for workflow automation and BI system.
B3.2: Business rules for automated disbursement of reports and templates on regular basis through external e-mail system.

B3.3: Migration tools to extract the data from an existing application source and feed the new system.

B4 Security

Example

B4.1: The software should have role-based access control. The user-profile-related configuration files should be maintained centrally on the server.
B4.2: The system should be LDAP compliant and support single sign-on if possible.
B4.3: The system should use encryption technologies and specific internet security mechanisms such as PKI, SSL etc. Accessibility should be through properly configured firewalls.

C: Operation

C1 Installation/Implementation

Example

C1.1: The vendor should use appropriate project management methodology and tools to implement the solution. The vendor should demonstrate the use of best practices while sizing the database, infrastructure etc. and the necessary tools to carry out the same.
C1.2: The vendor should be able to remotely login the test environment to fix any technical issues during and after the implementation. There should be a formal procedure for escalating and responding to the queries as per the predefined SLA.

C2 Services and Support

Example

C2.1: The vendor should continue to support (backward compatibility) the installed version for at least the period of the next five years or else provide the necessary upgrades without any charge. The vendor should implement new releases and upgrades to coincide with the changes in the dependant technologies and applications.
C2.2: The vendor should provide the necessary support 24/7 to close any technical or business functionality related issue arising after the expiry of the warranty period.

Once the business requirements are clearly defined, the decision to build or buy is taken according to the parameters such as time to market, availability of required competencies in-house, cost of development, cost of purchase and business value of the application. The decision whether to buy or build depends upon factors such as:

Longevity of the vendor: Check for the stability and the commitment of the vendor. If the product meets the business requirements, one should check if the product support from the vendor will continue to exist or not.

Time to implement: Estimate the business functionality, i.e. the size, of the software to be built. If the time to develop the solution meets the business deadline and the software requirements are stable and clear (i.e. the solution covers the business need holistically, with no foreseeable expansion in its scope), the solution could be built in-house.

In-house resources and competence: Assess the knowledge of the technology and the business domain available in-house to build the business solution, including the success rate of the development team. If the organisation is not equipped with adequate skills and knowledge to build the solution, it should opt for the buy decision to avoid risks.

Cost differential: If the cost differential between build and buy is too high, go for the option that has the lowest total cost of ownership. The cost includes cost of integration and ongoing maintenance apart from the acquisition costs.

Competitive advantage: If the investment in the project doesn't provide any boost to the core competencies, buying may be a good option.

Cost-benefits and functionality: If buying an existing solution helps the organisation to meet multiple business objectives (apart from the one in hand) at the cost of building one solution, the buying option is advisable.

Once the decision to buy the software package is taken, the procurement team prepares the list of potential vendors and solutions. Software directories, market survey reports, industry websites etc. can point to the names of potential vendors and products.

The business requirements converted into a formal RFP are finally forwarded to the potential vendors.

6.3 Evaluate Software Vendors and Package Solutions (Step 2)

Commercially viable solutions should be approached cautiously because (a) vendors may be new in market, (b) products may not be proven in the market, (c) architecture may be immature, (d) solutions might not be designed for flexibility and evolvability and (e) solutions might be using standards that are still evolving.

Before detailed evaluation of software vendors and solutions using AHP, an initial shortlist is prepared on the basis of certain filter criteria like cost, technology, vendor's market position etc. The idea is to eliminate the vendors and the products that differ widely in terms of the desired criteria and limit the detailed evaluation to only a few vendors and products.

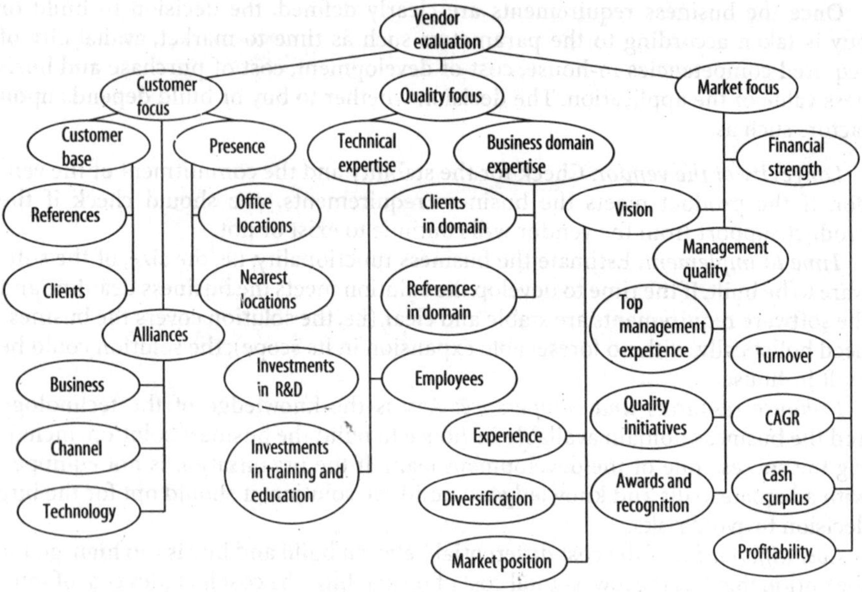

Figure 6.2 Hierarchy of vendor evaluation criteria.

The top two or three vendors are selected for detailed evaluation using AHP.

Vendor evaluation: Consider the following set of criteria for vendor evaluation:

1. Vision.
2. Financial stability.
3. Geographical presence.
4. Technical expertise.
5. Business domain focus.
6. Overall customer base and references.
7. Alliances.
8. Management quality.

Figure 6.2 shows the hierarchy of the vendor evaluation criteria. The evaluation criteria are compared and ranked relatively using the AHP (see Appendix 6.1).

The weights assigned to each of the criteria are summarised in Table 6.1.

The top two vendors (Vendor A and Vendor B) are ranked (see Appendix 6.1) on each of the criteria as listed in Table 6.1.

Table 6.2 summarises the relative rankings (on each of the criteria listed in Table 6.1) of Vendor A and Vendor B.

Software evaluation: Consider the following criteria for evaluating a software solution:

1. Business functionality.
2. Technology.
3. Operations.

Table 6.3 summarises the software evaluation criteria as described in Section 6.2.

Table 6.1 Criteria and corresponding weights for evaluation of vendors.

No.	*Criterion*	*Weight*	*Sub-criterion*	*Weight*
1	**Customer focus**	**0.258**		
1.1	Overall customer base and references	0.680	Clients	0.125
			References	0.875
1.2	Alliances	0.100	Business alliances	0.279
			Technology alliances	0.649
			Channel alliances	0.072
1.3	Geographic presence	0.220	Office locations	0.167
			Nearest locations	0.833
2	**Quality focus**	**0.637**		
2.1	Technical expertise	0.250	Employees	0.637
			Investments in R&D	0.258
			Investments in training and education	0.105
2.2	Business domain focus	0.750	Clients in the domain	0.125
			References in the domain	0.875
3	**Market focus**	**0.105**		
3.1	Vision	0.071	Experience	0.618
			Diversification	0.086
			Market position	0.297
3.2	Management quality	0.270	Top management experience	0.637
			Quality initiatives	0.258
			Awards and recognitions	0.105
3.3	Financial stability	0.656	Turnover	0.057
			CAGR	0.145
			Profitability	0.558
			Cash surplus	0.240

Table 6.2 Relative ranks of the vendors.

No.	*Sub-criterion*	*Vendor A*	*Vendor B*
1	Clients	0.750	0.250
	References	0.875	0.125
2	Business alliances	0.167	0.833
	Technology alliances	0.250	0.750
	Channel alliances	0.125	0.875
3	Office locations	0.833	0.167
	Nearest locations	0.750	0.250
4	Employees	0.875	0.125
	Investments in R&D	0.167	0.833
	Investments in training and education	0.250	0.750
5	Clients in the domain	0.125	0.875
	References in the domain	0.833	0.167
6	Experience	0.833	0.167
	Diversification	0.750	0.250
	Market position	0.875	0.125
7	Top management experience	0.833	0.167
	Quality initiatives	0.750	0.250
	Awards and recognitions	0.875	0.125
8	Turnover	0.167	0.833
	CAGR	0.250	0.750
	Profitability	0.125	0.875
	Cash surplus	0.167	0.833

Table 6.3 Criteria for evaluation of software products.

No.	Criterion	Weight	Sub-criterion (1)	Weight	Sub-criterion (2)	Weight	Sub-criterion (3)	Weight
1	Business functionality	0.68	A1	0.051				
			A2	0.145				
			A3	0.558				
			A4	0.240				
2	Technology	0.10	B1	0.064	B1.1	0.68	B1.1.1	0.637
							B1.1.2	0.258
							B1.1.3	0.105
					B1.2	0.10	B1.2.1	0.833
							B1.2.2	0.167
					B1.3	0.22	B1.3.1	0.637
							B1.3.2	0.258
							B1.3.3	0.105
			B2	0.124	B2.1	0.064	B2.1.1	0.833
							B2.1.2	0.167
					B2.2	0.122	B2.2.1	0.750
							B2.2.2	0.250
					B2.3	0.572	B2.3.1	0.875
							B2.3.2	0.125
					B2.4	0.240	B2.4.1	0.167
							B2.4.2	0.833
			B3	0.572	B3.1	0.071		
					B3.2	0.233		
					B3.3	0.656		
			B4	0.240	B4.1	0.618		
					B4.2	0.086		
					B4.3	0.297		
3	Operations	0.22	C1		C1.1	0.250		
					C1.2	0.750		
			C2		C2.1	0.125		
					C2.2	0.875		

The products are compared and ranked relatively (see Appendix 6.1) on each of the criteria listed in Table 6.3.

Table 6.4 summarises the ranking of the products (Product A from Vendor A and Product B from Vendor B).

Using the above set of evaluation criteria, the combined scorecard for ranking the alternatives is:

$$\text{Combined rating} = \phi_1\,(\text{Vendor rating}) + \phi_2\,(\text{Product rating})$$

where ϕ_1, ϕ_2 are weights assigned to vendor and product based on their relative importance (differs case to case) and $\phi_1 + \phi_2 = 1$ (see Table 6.5).

From Table 6.5:

$$\phi_1 = 0.250, \ \phi_2 = 0.750$$

Vendor rating: The overall ranking of the vendors (using Table 6.2), is:

Vendor A: 0.696, Vendor B: 0.293

Product rating: The overall ranking of the products (using Table 6.4), is:

Product A: 0.752, Product B: 0.462

Table 6.4 Product rankings and the criteria for evaluation of software products.

No.	*Criterion*	*Sub-criterion (1)*	*Sub-criterion (2)*	*Sub-criterion (3)*	*Product A*	*Product B*
1	Business functionality	A1			0.25	0.75
		A2			0.125	0.875
		A3			0.833	0.167
		A4			0.75	0.25
2	Technology	B1	B1.1	B1.1.1	0.875	0.125
				B1.1.2	0.167	0.833
				B1.1.3	0.25	0.75
			B1.2	B1.2.1	0.125	0.875
				B1.2.2	0.833	0.167
			B1.3	B1.3.1	0.75	0.250
				B1.3.2	0.875	0.125
				B1.3.3	0.167	0.833
		B2	B2.1	B2.1.1	0.167	0.833
				B2.1.2	0.25	0.750
			B2.2	B2.2.1	0.125	0.875
				B2.2.2	0.167	0.833
			B2.3	B2.3.1	0.25	0.750
				B2.3.2	0.167	0.833
			B2.4	B2.4.1	0.25	0.750
				B2.4.2	0.125	0.875
		B3	B3.1		0.833	0.167
			B3.2		0.75	0.25
			B3.3		0.875	0.125
		B4	B4.1		0.833	0.167
			B4.2		0.25	0.750
			B4.3		0.125	0.875
3	Operations	C1	C1.1		0.833	0.167
			C1.2		0.75	0.25
		C2	C2.1		0.875	0.125
			C2.2		0.167	0.833

Table 6.5 Weights assigned to vendor and product.

	Vendor	*Product*	*Eigenvector*
Vendor	1	1/3	0.250
Product	3	1	0.750

Combined ranking of products: The combined ranking of the products, using Table 6.5, is:

$$\text{Product A: } 0.25 \times (0.696) + 0.75 \times (0.752) = 0.738$$

$$\text{Product B: } 0.25 \times (0.293) + 0.75 \times (0.462) = 0.420$$

Product A scores more than Product B in combined ranking. Hence product A best suits the business-, technology- and operations-related requirements of the organisation.

Detailed due diligence: The project team should perform a detailed due diligence to validate the rankings achieved through the previous steps by one or more of the following approaches: (1) product documentation, (2) reference site visits, (3) product demonstration, (4) trial installation etc.

6.4 Negotiation (Step 3)

The vendors who are finally selected after detailed due diligence are notified for negotiations. The negotiation deals with various issues such as acquisition cost, payment terms, maintenance, licence fees, post-installation support, service levels, implementation, and legal terms etc.

The following is an illustrative approach to contract negotiation:

1. Make a list of elements that are negotiable and assign the criticality and goal to it. Example: Table 6.6.
2. Use information gathered during the detailed due diligence on items such as reference site visits, market trends and competition to set the goal for discounts.
3. Analyse the negotiable elements and their impact on the budget for the current and next five years. For instance, future discounts help in controlling the expenses that may occur in future for reasons such as the underestimation of the number of licences in the beginning.
4. Analyse the specific and critical requirements such as 24-hour service support, remote login and troubleshoot, or the specific minimum performance level as expected, and address them in the warranties. The procurement team should clearly outline business-critical requirements and cap them under the warranties, which if not fulfilled by the vendor should invite penalties or refunds or discounted services etc.
5. Analyse the impact of the applications on supplementary purchases or reusing the existing infrastructure such as the hardware, operating system, databases etc. Use the study to bargain and negotiate a better deal.

The procurement team should form a negotiation team and pursue the above-mentioned practices to optimise the deal. The negotiation team should have the following roles:

1. A single vendor interface (if needed, have an additional role that can keep the discussion going while the first one tries to negotiate hard with the vendor).
2. A researcher, who collates the information about the vendors through multiple sources. The researcher's role includes studying the vendor, its customers and the deals in the past. The researcher feeds the output to the entire negotiation team.
3. A financial analyst who helps in predicting the impact of the purchase on the current and next five fiscal years. This includes creating multiple scenarios and measuring the financial risk involved, apart from helping the team to come out with the business case.

Table 6.6 Example of negotiation elements and illustrative goals.

Negotiable elements	*Criticality*	*Goal*
List or acquisition price	High	Reduction by 20–30%
Maintenance and support costs	High	Reduction by 10–20%, 2–5% maximum cap for year-to-year increase in maintenance costs
Implementation costs	Low	Reduction by 20–30%
Future discounts	Medium	30–40% discounts on purchases in the next 3 years, continued training support during the life of the contract etc.
Service levels	High	Refund, discount or penalties if service level agreements are not obeyed

References

1. Dickson GW (1966) An analysis of vendor selection systems and decisions. Journal of Purchasing 2(1): 5–17.
2. Stele P, Court B (1996) Profitable Purchasing Strategies. McGraw-Hill.
3. Lee L Jr, Dobler DW (1977) Purchasing and Materials Management. TMH Edition.
4. Kuo K, Wilson N (2001) The scientific art of contract negotiations. Educate Quarterly (1).

Appendix 6.1

A: Vendor Evaluation

Vendor Evaluation Criteria

	Market focus	*Customer focus*	*Quality focus*
Market focus	1	1/3	1/5
Customer focus	3	1	1/3
Quality focus	5	3	1
N.E.V.	0.105	0.258	0.637

Customer Focus

	Customer base (references)	*Alliances*	*Geographic presence*
Customer base (references)	1	3	7
Alliances	1/3	1	1/5
Geographic presence	1/7	5	1
N.E.V.	0.680	0.100	0.220

Market Focus

	Vision	*Management quality*	*Financial stability*
Vision	1	1/5	1/7
Management quality	5	1	1/3
Financial stability	7	3	1
N.E.V.	0.071	0.273	0.656

Quality Focus

	Technical expertise	*Business domain expertise*
Technical expertise	1	1/3
Business domain expertise	3	1
N.E.V.	0.250	0.750

Overall Customer Base and References

	Clients	*References*
Clients	1	1/7
References	7	1
N.E.V.	0.125	0.875

Vision

	Experience	*Diversification*	*Market position*
Experience	1	5	3
Diversification	1/5	1	1/5
Market position	1/3	5	1
N.E.V.	0.618	0.086	0.297

Financial Stability

	Turnover	*CAGR*	*Profitability*	*Cash surplus*
Turnover	1	1/5	1/7	1/3
CAGR	5	1	1/5	1/3
Profitability	7	5	1	3
Cash surplus	3	3	1/3	1
N.E.V.	0.057	0.145	0.558	0.240

Quality Focus

	Office locations	*Nearest locations*
Office locations	1	1/5
Nearest locations	5	1
N.E.V.	0.167	0.833

Business Domain Focus

	Clients in domain	*References in domain*
Clients in domain	1	1/7
References in domain	7	1
N.E.V.	0.125	0.875

Alliances

	Business	*Technology*	*Channel*
Business	1	1/3	5
Technology	3	1	7
Channel	1/5	1/7	1
N.E.V.	0.279	0.649	0.072

Technical Expertise

	Employees	*Investments in R&D*	*Investments in training and education*
Employees	1	5	3
Investments in R&D	1/5	1	5
Investments in training and education	1/3	1/5	1
N.E.V.	0.637	0.258	0.105

Management Quality

	Top management experience	*Quality initiatives*	*Awards and recognitions*
Top management experience	1	5	3
Quality initiatives	1/5	1	5
Awards and recognitions	1/3	1/5	1
N.E.V.	0.637	0.258	0.105

Experience, Office Locations, Reference in Domain, Top Management Experience

	Vendor A	*Vendor B*	*N.E.V.*
Vendor A	1	5	0.833
Vendor B	1/5	1	0.167

Diversification, Nearest Locations, Clients, Quality Initiatives

	Vendor A	*Vendor B*	*N.E.V.*
Vendor A	1	3	0.750
Vendor B	1/3	1	0.250

Market Position, Employees, References, Awards and Recognition

	Vendor A	*Vendor B*	*N.E.V.*
Vendor A	1	7	0.875
Vendor B	1/7	1	0.125

Turnover, Cash Surplus, Investment in R&D, Business Alliance

	Vendor A	*Vendor B*	*N.E.V.*
Vendor A	1	1/5	0.167
Vendor B	5	1	0.833

CAGR, investments in Training and Education, Technology Alliance

	Vendor A	*Vendor B*	*N.E.V.*
Vendor A	1	1/3	0.250
Vendor B	3	1	0.750

Profitability, Clients in Domain, Channel Alliance

	Vendor A	*Vendor B*	*N.E.V.*
Vendor A	1	1/7	0.125
Vendor B	7	1	0.875

B: Package Evaluation

Software Evaluation Criteria

	Business functionality	*Technology*	*Operations*
Business functionality	1	3	7
Technology	1/3	1	1/5
Operations	1/7	5	1
N.E.V.	0.680	0.100	0.220

Business Functionality

	A1	A2	A3	A4
A1	1	1/5	1/7	1/3
A2	5	1	1/5	1/3
A3	7	5	1	3
A4	3	3	1/3	1
N.E.V.	0.057	0.145	0.558	0.240

Technology

	B1	B2	B3	B4
B1	1	1/3	1/7	1/3
B2	3	1	1/5	1/3
B3	7	5	1	3
B4	3	3	1/3	1
N.E.V.	0.064	0.124	0.572	0.240

B1 Usability

	B1.1	B1.2	B1.3
B1.1	1	3	7
B1.2	1/3	1	1/5
B1.3	1/7	5	1
N.E.V.	0.680	0.100	0.220

B1.1 Personalisation

	B1.1.1	B1.1.2	B1.1.3
B1.1.1	1	5	3
B1.1.2	1/5	1	5
B1.1.3	1/3	1/5	1
N.E.V.	0.637	0.258	0.105

B1.2 Accessibility

	B1.2.1	B1.2.2	N.E.V.
B1.2.1	1	5	0.833
B1.2.2	1/5	1	0.167

B1.3 Learnability

	B1.3.1	*B1.3.2*	*B1.3.3*
B1.3.1	1	5	3
B1.3.2	1/5	1	5
B1.3.3	1/3	1/5	1
N.E.V.	0.637	0.258	0.105

B2 Architecture

	B2.1	*B2.2*	*B2.3*	*B2.4*
B2.1	1	1/3	1/7	1/3
B2.2	3	1	1/5	1/3
B2.3	7	5	1	3
B2.4	3	3	1/3	1
N.E.V.	0.064	0.124	0.572	0.240

B2.1 Architectural Compliance

	B2.1.1	*B2.1.2*	*N.E.V.*
B2.1.1	1	5	0.833
B2.1.2	1/5	1	0.167

B2.2 Scalability

	B2.2.1	*B2.2.2*	*N.E.V.*
B2.2.1	1	3	0.750
B2.2.2	1/3	1	0.250

B2.3 Interoperability

	B2.3.1	*B2.3.2*	*N.E.V.*
B2.3.1	1	7	0.875
B2.3.2	1/7	1	0.125

B2.4 Performance

	B2.4.1	*B2.4.2*	*N.E.V.*
B2.4.1	1	1/5	0.167
B2.4.2	5	1	0.833

B3 Integration

	B3.1	B3.2	B3.3
B3.1	1	1/5	1/7
B3.2	5	1	1/3
B3.3	7	3	1
N.E.V.	0.071	0.273	0.656

B4 Security

	B4.1	B4.2	B4.3
B4.1	1	5	3
B4.2	1/5	1	1/5
B4.3	1/3	5	1
N.E.V.	0.618	0.086	0.297

C1 Installation/Implementation

	C1.1	C1.2
C1.1	1	1/3
C1.2	3	1
N.E.V.	0.250	0.750

C2 Services and Support

	C2.1	C2.2
C2.1	1	1/7
C2.2	7	1
N.E.V.	0.125	0.875

A1, B1.1.3, B2.1.2, B2.3.1, B2.4.1, B4.2

	Product A	Product B	N.E.V.
Product A	1	1/3	0.250
Product B	3	1	0.750

A2, B1.2.1, B2.2.1, B2.4.2, B4.3

	Product A	Product B	N.E.V.
Product A	1	1/7	0.125
Product B	7	1	0.875

A3, B1.2.1, B3.1, B4.1, C1.1

	Product A	*Product B*	*N.E.V.*
Product A	1	5	0.833
Product B	1/5	1	0.167

A4, B1.3.1, B3.2, C1.2

	Product A	*Product B*	*N.E.V.*
Product A	1	3	0.750
Product B	1/3	1	0.250

B1.1.1, B1.3.2, B3.3, C2.1

	Product A	*Product B*	*N.E.V.*
Product A	1	7	0.875
Product B	1/7	1	0.125

B1.1.2, B1.3.3, B2.1.1, B2.2.2, B2.3.2, C2.2

	Product A	*Product B*	*N.E.V.*
Product A	1	1/5	0.167
Product B	5	1	0.833

7 Estimating the Software Application Development Effort at the Proposal Stage

7.1 Early Software Effort Estimation and Techniques

A company that won a $350 000 software development project came to know that the next lowest bid by the competitor was $1.4 million. A few weeks later, after the requirement engineering phase, the company finds that the estimate during the proposal generation was highly inaccurate and underestimated. At the same time, if the bid had been overestimated, the company wouldn't have got the contract at all.

Such wide variations in price are common and are due to inaccurate cost estimation at an early stage of development such as a proposal submission or bidding stage. At this stage, not much information is known about the software application to be developed. The existing cost estimation models can provide an accurate estimate only when the requirements are completely specified. Under such circumstances, the pricing decisions are based on the cost estimation using analogies and the opinions of experts.

Most companies lack a structured and disciplined approach in applying these heuristic techniques. There is a need to have a structured cost estimation decision-making process that can remove the inconsistencies and biases besides helping to retain the expert knowledge gained from each such iteration.

One of the approaches practised is the use of the AHP to evaluate expert opinions and to rank the historical projects in order to draw useful analogies. The organisation should combine this with the practice of building a repository of reference models or the use-case catalogues.

Total software cost also includes costs pertaining to effort, insurance, travel, training and overheads (facilities, infrastructure etc.). Estimating the software effort accurately is the most critical part of the cost estimation process. Software effort is measured in person-months (PM) or person-days and is defined as the accumulation of the amount of time spent by each person over the life cycle of the project to develop the software. The total software effort in a project includes the effort in management activities such as project planning, tracking, scheduling, staffing and reviewing.

Estimating software effort at an early stage, when no data about the application to be developed are available, is driven primarily by heuristics.

The heuristics estimation includes techniques such as:

1. *Expert opinion*: The estimate is arrived at using the joint opinion of a group of experts based on their past experience. The experts use their respective

experiences to exchange predictions of the software costs until a consensus is reached. This is a relatively inexpensive estimation method and can be accurate if and only if the experts have experience of similar kinds of projects. Under exceptional circumstances, such as the non-availability of historical data and new projects (i.e. which were never executed before), expert opinion becomes the only choice for estimation. The disadvantages include personal bias, incomplete recall, and the unaccountability of the extent of knowledge (leveraged/applied by experts) etc.

2. *Analogy*: The estimate is arrived at by comparing the system to be developed with previous projects which belong to the same application domain. It can be as accurate as the accuracy of the historical data and the similarity among the applications.
3. *Bottom-up*: The estimate is achieved using the sum of estimates of its components derived individually, i.e. the estimation begins at the subsystem level and these estimates are added to get a final estimate. This approach can be used if details like architecture and design (i.e. the component level details) of the system are known.
4. *Top-down*: The estimate is derived using global properties at system level, which is then divided among the individual components, i.e. the estimation begins at the system level to evaluate overall system functionality and is then broken down into individual sub-system levels. This can underestimate the effort owing to the unidentified nature of technical difficulties or issues at sub-system levels.
5. *Price-to-win*: The estimate is tuned or manipulated in order to suit the price to win a contract. The price limits the size of the software functionality to be programmed and reduces the probability of delivering complete software. This often leads to a situation of conflict and iterative negotiations to extend the price to cover the costs.
6. *Parkinson's law*: The estimate is derived assuming that the work will expand to fill the volume, i.e. will completely consume the available capacities/resources to develop the software. Despite the maximisation of the utilisation of the available resources, the project may fall short of the required resources by huge margins.

The cost estimation is performed continuously across the various phases of the software development life cycle (i.e. requirements, design, coding, testing etc.). Once the requirements are completely specified, algorithmic methods can be applied in order to achieve objective, repeatable, analysable and accurate cost estimates. Unlike heuristics, algorithmic methods employ a parametric function that uses multiple parameters associated with the software product, e.g. development process, development personnel, project infrastructure etc., as an input to arrive at an estimate.

These functions may be defined based on observation, regression or mathematical formulae and are constantly refined with insights gained from previous projects and research (example COCOMO II).

There are complex methods that are being used to estimate cost/effort like learning (example: neural networks and case-based reasoning) and simulation (example: system dynamics) based approaches.

For the sake of simplicity, Figure 7.1 broadly classifies the algorithmic methods into two categories as follows:

A. *Dynamic*: For the purposes of estimation the behaviour of the development environment (resources, workload, requirements, productivity etc.) is assumed

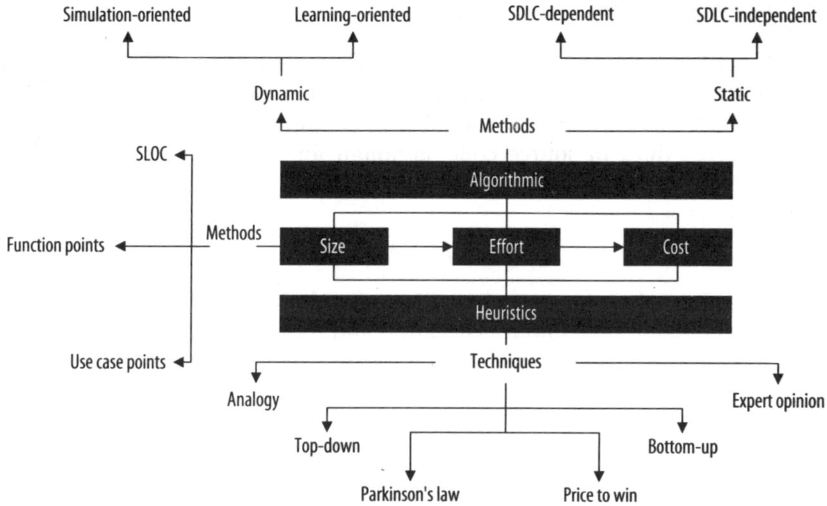

Figure 7.1 Classification of software estimation methods.

to be dynamic and the model (e.g. a simulation model) allows the estimator to build various project scenarios and produce estimates for them. In some cases (e.g. that of a learning-oriented model), the model is designed to self-adjust its variables iteratively over a period of its usage so as to improve the accuracy of estimation. Compared with the static ones, these models are relatively complex to operate.

B. *Static*: During the estimation, the behaviour of the development environment is considered to be static, i.e. the total effort estimate always remains the same irrespective of any constraints. But the model may provide an estimate for the effort that is either independent of the duration of the software development life cycle (SDLC) (e.g. COCOMO II) or else dependent on the time and/or stages in development (i.e. the effort varies with the time elapsed in the SDLC) (e.g. the SLIM Model).

As depicted in Figure 7.1, the software size has to be estimated prior to estimating the software effort. The most widely used software size estimation methods are described in Section 7.2. Section 7.3 describes a systematic approach to estimating the size using the estimation workbench and the AHP. (See Appendix 7.1.)

7.2 Software Size and Effort Estimation Methods

Size is a measure of the amount of software to be developed to satisfy the requirements and could be estimated using methods such as lines of code (LOC), function points and use case points. There are various other methods (like feature points) that either are variants of function points or are similar in the sense that they also measure functional size but using different components like screens, reports etc., as in the case of object points.

7.2.1 Lines of Code (LOC)

This method has been used extensively ever since and much before the formal study of software estimation techniques began in the 1960s. The total length of the code is calculated as the sum of the number of commented (CLOC) and non-commented (NLOC) lines in source code, although for all practical purposes only NLOC is considered to be the software size. KLOC represent size in thousands of lines of code.

$$LOC = NCLOC + CLOC$$

The subjective values for optimistic lines of code (minimum LOC, O_{LOC}), pessimistic lines of code (maximum LOC, P_{LOC}) and most likely lines of code (mean LOC, M_{LOC}) are also used to estimate the software size in LOC as follows:

$$LOC = \frac{1}{6}\,(O_{LOC} + 4M_{LOC} + P_{LOC})$$

$$\sigma = \frac{1}{6}\,(P_{LOC} - O_{LOC})$$

The above estimation is based on the calculation of the mean (average) estimation of the stochastically distributed activities of the project used in PERT.

7.2.2 Function Point (FP)

Developed in 1979 by Allan J. Albrecht of IBM, function point is a method that measures the software size as a magnitude of the functionality delivered by the system to its users. Function point estimation is based on the count of five basic component types of the system:

1. *External interface files (EIF)*: The logically related data or control information needed by the application and maintained within the boundary of an external application.
2. *Internal logical files (ILFs)*: The logically related data or control information needed by the application and maintained within its own boundary.
3. *External inputs (EI)*: The processes pertaining to the incoming data and control information coming from outside the boundary of the application.
4. *External outputs (EO)*: The processes pertaining to the outgoing data and control information (using derivations, calculations etc.) and sends across the boundary of the application. The process may require maintaining one or more ILFs to alter the system behaviour.
5. *External inquiries (EQ)*: The processes that retrieve the data and control information and sends across the boundary of the application. These processes neither maintain any ILFs nor alter the behaviour of the system.

Component types (1) and (2) are known as data functions (DF) and they col-lectively represent the functionality corresponding to the external and internal data needed by an application. Component types (3), (4) and (5) are known as transactional functions (TF) and represent collectively the functionality

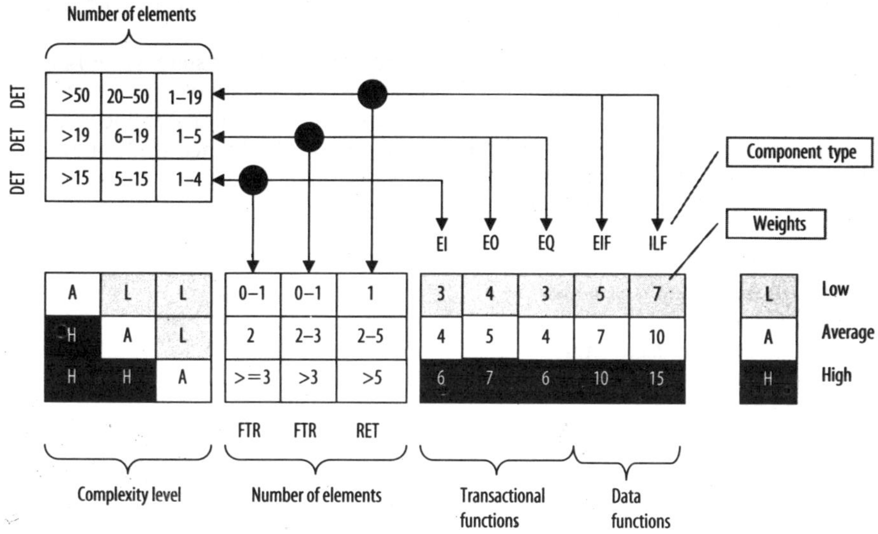

Figure 7.2 Function point analysis: component types, elements and complexity table.

corresponding to the procedures that process the data as needed by the user. The complexity of a component type is based on the number of elements that constitute it.

The complexity of a DF component (i.e. an EIF or ILF) is based on the number of:

1. *Data element types (DET)*: Non-repeated fields as recognised by the user and maintained in ILFs or EIFs.
2. *Record element types (RET)*: Subgroups (both optional and mandatory) of data elements in ILFs or EIFs.

The complexity of a TF component (i.e. an EI, EO or EQ) is based on the number of:

1. *File type references (FTR)*: EIFs accessed or ILFs accessed/maintained by the process, and
2. *Data element types (DET)*: non-repeated fields as recognised by the user and maintained in or retrieved from ILFs or EIFs by the application/ process.

There are objective guidelines and complexity tables that help in finding the level of complexity (low (L), average (A), high (H)) of a component type and assigning the corresponding numerical weight (see Figure 7.2). For instance, an ILF with 4 RET and 30 DET will have "average" complexity and a numeric weight of 10.

The software size as a sum of the weights of all the components is expressed in unadjusted function points (UFP) as:

$$UFP = \sum_{i=1}^{5} \sum_{j=1}^{3} C_{ij} W_{ij}$$

where C_{ij} is the number and W_{ij} is the weight of a component of type i with a level of complexity j.

UFP is adjusted to account for the overall complexity of the system using 14 general characteristics of the system (GCS), measured on a scale of 0 to 5 (0 = not present or no influence, 1 = incidental influence, 2 = moderate influence, 3 = average influence, 4 = significant influence, 5 = strong influence throughout) to obtain final function points (FP).

The 14 general characteristics are:

- Data communication.
- Distributed data processing.
- Performance.
- Heavily used configuration.
- Transaction rate.
- Online data entry.
- End user efficiency.
- Online updating.
- Complex processing.
- Reusability.
- Installation ease.
- Operational ease.
- Multiple sites.
- Facilitation of change.

The above 14 characteristics of the system are used as input to adjust the size of the software in UFP using a value adjustment factor (VAF). This adjusts the UFP by at most ±35%.

VAF is computed by adding the numeric weight corresponding to each of the GSCs:

$$VAF = 0.65 + 0.01\sum_{i=1}^{14} V_i$$

where V_i is the value of the ith GSC and
Finally, the software size in function points is calculated as:

$$FP = UFP \times VAF$$

Example

An EI with 1 FTR and 24 DET elements will have a high complexity, thus a numeric weight of 6.

7.2.3 Use Case Points (UCP)

Developed in 1993 by Gustav Karner of Objectory (now Rational Software), use case points (UCP), like function points, measure software size in terms of the magnitude of the functionality delivered by the system to its users.

UCP count is based on the number and complexity of two component types, (1) actors and (2) use cases.

A use case is a set of scenarios or use case instances that reflect the sequence of actions performed by the system to deliver a meaningful result to an actor.

Unadjusted use case points (UUCP) are calculated as follows:

$$UUCP = \sum_{i=1}^{3} A_i W_i + \sum_{j=1}^{3} U_j X_j$$

where A_i is the number and W_i the weight of actors with level of complexity i, and U_j is the number and X_j the weight of use cases with level of complexity j (see Figure 7.3).

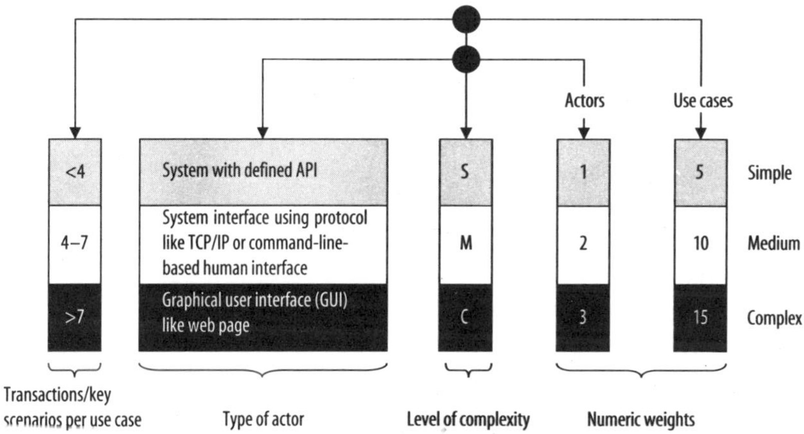

Figure 7.3 Use case point analysis: actors, transactions and use case complexity table.

An actor is a user of a system and is ranked as one of the following (based on its type):

1. Simple (S) [$W_1 = 1$]: an external system with a predefined application programming interface (API).
2. Medium (M) [$W_2 = 2$]: an external system interface using a protocol such as TCP/IP or a command line human interface.
3. Complex (C) [$W_3 = 3$]: graphical-user-interface- (GUI) based human interaction.

Use case is ranked as one of the following (based on the number of key scenarios or transactions supported):

1. Simple (S) [$X_1 = 5$]: a use case supporting 3 or fewer key scenarios.
2. Medium (M) [$X_2 = 10$]: a use case supporting 4 to 7 key scenarios.
3. Complex (C) [$X_3 = 15$]: a use case supporting more than 7 key scenarios.

UUCP is then adjusted to account for the overall complexity of the system to arrive at size in use case points:

$$UCP = UUCP \times (TCF = 0.6 + (0.01)\sum_{i=1}^{13} T_i Y_i)$$
$$\times (EF = 1.4 + (-0.03) \times \sum_{j=1}^{8} E_j Z_j)$$

where T_i is the degree of relevance and Y_i is the weight of the technical complexity factor i, E_j is the degree of relevance and Z_j is the weight of the environment factor j. T_i and E_j are measured on a scale of 0 (= irrelevant or no influence) to 5 (= essential or strong influence).

The 13 technical complexity factors (TCF) and their corresponding weights are:

1. Distributedness of system [$Y_1 = 2$].
2. Response or throughput performance objectives [$Y_2 = 2$].
3. End-user efficiency [$Y_3 = 1$].
4. Complexity of processing [$Y_4 = 1$].
5. Code reusability [$Y_5 = 1$].
6. Ease of installation [$Y_6 = 0.5$].
7. Ease of use [$Y_7 = 0.5$].
8. Portability [$Y_8 = 2$].
9. Ease of change [$Y_9 = 1$].
10. Concurrentness [$Y_{10} = 1$].
11. Security features [$Y_{11} = 1$].
12. Accessibility for third parties [$Y_{12} = 1$].
13. Need for special training [$Y_{13} = 1$].

The 8 environment factors (EF) and their corresponding weights are:

1. Familiarity with the process [$Z_1 = 1.5$].
2. Application experience [$Z_2 = 0.5$].
3. Object-oriented experience [$Z_3 = 1$].
4. Lead analyst capability [$Z_4 = 0.5$].
5. Motivation [$Z_5 = 1$].
6. Stability of requirements [$Z_6 = 2$].
7. Part-time workers [$Z_7 = -1$].
8. Difficult programming language [$Z_8 = 2$].

7.2.4 Comparative Remarks

Accuracy in estimating the software size at the proposal stage is a key to cost estimation. The above-mentioned measures of software size should be looked at from the perspective of their predictive capabilities. Among all of them the one with the least predictive ability is LOC. The software size in LOC can be estimated with reasonable accuracy only on completion of the design or code development. An early estimate of size in LOC is difficult. Some of the other demerits of LOC include:

1. Dependency on the programming language; for example, the number of lines of code required to code a function in object-oriented language differs widely from the number of lines of code using procedural language.
2. Lack of a fixed standard or guidelines to measure the size in LOC; for example, every firm sets its own guidelines around the inclusion/exclusion of commented lines, non-commented lines, executed lines, non-executed lines, program statement or logical delimited, physical lines, new or modified or reused lines etc.

This is why most of the effort estimation parametric models (of the 1960s to 70s) based on size in LOC are not practical for prediction or estimation purposes. They are measures of result, computed after the completion of a project or a phase.

As compared with LOC, FP has better support for the early prediction of software size. Despite some of the criticisms that remain around (1) the application user's view-based size estimation, (2) the classification of the complexity level as low, average or high, (3) the choice of relative weights for these complexity levels and (4) the restriction of adjustment - limited to 14 GSCs - etc., this method is the most widely accepted and practised in the industry. Size in FP has the following advantages:

1. It is independent of the implementation technology, i.e. the programming language, hardware platform and development methodology.
2. Function point analysis is based on the logical design of the system from the user's perspective. This motivates non-technical estimators for participation in size estimation.

UCP is similar to FP but unlike FP, the UCP method is built on use cases that are highly subjective inputs. There are various points of debate:

1. *Key scenarios:* It is difficult to define a standard to identify key scenarios.
2. *Number of use cases:* It is difficult to define the granularity of a use case. The number of use cases depends upon the level of granularity while capturing the requirements.
3. *Actors:* The treatment of the relative ranks/weights for the complexity of the actors is open to debate, especially when inherent underlying

support from programming environments (such as .NET and J2EE) is available. Some of the UCP estimators may choose to ignore the points corresponding to the complexity of the actors.

Today, function points and use case points are widely used for reaching size estimates. The use of software size in FP and UCP along with heuristic techniques such as analogy and expert opinion for early prediction can still be challenged, as it is highly dependent upon the experience of estimators in applying heuristic techniques, the volume & accuracy of historical data and the maturity of estimation process within the organisation.

7.3 A Systematic Approach to Early Software Estimation

Early software estimation implies estimating the software effort/cost during the proposal or bidding process (see Figure 7.4) when not much is known in detail

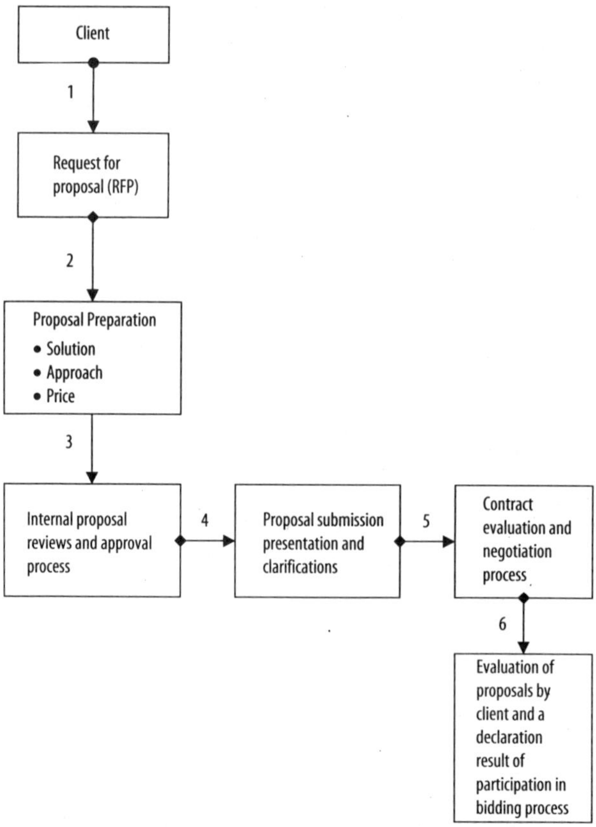

Figure 7.4 A typical RFP proposal process.

about the software product to be developed. Under such circumstances, an organisation depends heavily upon its experts and on historical data about similar projects. The most feasible approaches under such circumstances have been variants of Delphi and Analogy.

In the Delphi approach, a group of experts is asked to provide estimates. The estimates are collected and tabulated by the co-coordinator to find if the estimates are within an acceptable deviation from the group average (or within an acceptable range). In case the estimates are not in sync, revised estimates are asked for after distributing the tabulated estimates to the estimators, highlighting the personal estimate and keeping others as anonymous. The process continues till a consensus is reached. This approach is highly subjective and unstructured. Moreover, from the onset, the experts are not being empowered or disciplined to make use of the existing knowledge pool (the data on previous projects) in the organisation to improve their estimates.

In analogy-based estimation, the project database is searched to find out the metrics of the projects similar to the one to be developed. In this approach, the accuracy of the estimation is limited by the availability and effectiveness of the search of historical data for similar projects.

One can further improve the accuracy of the estimate if both the expert and the analogy-based estimation techniques are applied in tandem and the above-mentioned inherent limitations in both are overcome by applying the AHP.

The systematic approach being described here includes a recommendation for adopting the practice of using the following:

1. A framework to dynamically rank and search the projects in the databases using AHP.
2. Reference domain meta-models and reference use case catalogues based on patterns. The collection is enriched over a period of time and retains the knowledge of an expert applied during an instance of estimation for a project.
3. A framework to dynamically rank the experts based on their competencies, past performances (i.e. accuracy in estimation) and relevant experiences.

A reference domain meta-model is a model of an application in a particular business or application domain. The reference meta-model based on the fundamentals of function point analysis can help an organisation in capturing the business requirements using five component types. The model can be classified for a particular application domain for a business vertical.

The organisation can create a database to store and categorize the software requirements into EI, EO, EQ, ILF and EIF, with corresponding details of DET, FTR and RET. The project teams on completion of their projects can use the same database to link the project metrics (actual and estimated) such as effort, software size and team size etc. to the model constructed in the requirement stage.

Such a practice, over a period of time, will help categorize the new project and also help its component types to map to pre-existing identified patterns such as "report generation", "control information", "screen navigation", "file export" etc. and reuse the FP estimation and actual metrics from other projects executed in the past.

The database can also be used to link the logical design of the application to the underlying implementation constructs such as component design or source code. This will help in the optimisation of the development time and effort by suitably leveraging the benefits from reusability. Reusability benefits are further enhanced if the organisation has implemented a standardised approach to software development practices like the use of requirement patterns, object modelling, business modelling, data modelling, architecture modelling, component-based development and model-driven architecture.

An organisation that employs use cases to capture business requirements can through their systematic collection build a reference use case catalogue categorised around patterns that have direct traceability to implementation constructs.

As an example, we will assume that an organisation has a practice of use-case-based requirement gathering and estimation. The company using the UCP method to estimate size can expect the reference use case catalogue to provide an initial list of possible use cases applicable to the software to be developed. The richness of the reference use case catalogue depends upon the following:

1. The maturity of the estimation practice adopted in the organisation.
2. An estimation workbench that helps store, classify and search the project history, apart from automating the maintenance and creation of use case catalogues.
3. A knowledge of the business and application domains within the enterprise

Figure 7.5 depicts the systematic approach as a three-step process.

Assuming that an organisation follows use-case-based requirement collection and estimation practices, the activities performed in each step will be as in the following summary:

Step 1: This step is performed by business domain expert(s). It is comprises the following activities:

(a) Read the RFP carefully.
(b) Use the information provided in the RFP and look for similar requirement patterns and use cases implemented in previous projects. If possible, leverage industry-specific reference frameworks for applications and business processes.
(c) Complement information gained in A and B to arrive at a minimal set of use cases.

Step 2: This step is performed by application domain experts. It comprises the following activities:

(a) Defining the factors (in hierarchical manner) that contribute to the complexity and risk of the project. The organisation may decide to use (the subset of) the TCF and EF for this purpose. The current project/system is ranked (using the AHP) according to these factors.
(b) Using the current project/system attributes to form the search criteria to draw analogies with similar projects executed in the past. Search criteria help in drawing analogies on two aspects: (i) similarity in technology, functionality and process, and (ii) similarity in the complexity and risk being faced by the team. A cutoff on the extent of similarity (i.e. differences between ranks) may be used to limit the projects in order draw analogies.
(c) Extending the use case set as produced by the business domain expert in Step 1, based on the analogies and a brainstorming exercise.
(d) Updating the reference use catalogue, if the brainstorming exercise has resulted in identification of a new requirement pattern.
(e) Using the AHP, evaluating the application expert's estimates on the basis of the factors that affect the accuracy of the expert opinion.
(f) Calculating the final size in UUCP on the basis of the extended use case set and the evaluated expert opinions.

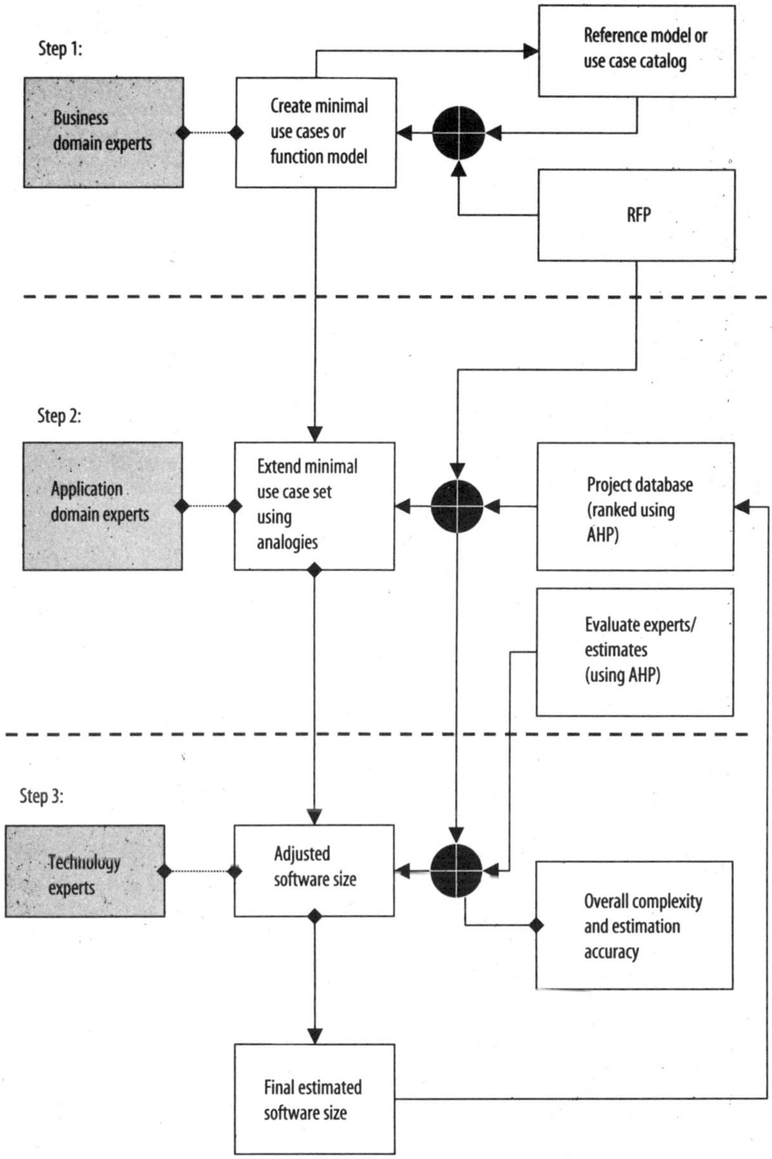

Figure 7.5 Systematic approach to early software estimation.

Step 3: This step is performed by technology domain expert(s). It consists of the following activities:

(a) Adjusting the UUCP using the 13 TCF and the 8 EF factors to arrive at the following software size in UCP.
(b) Identifying the relative errors in the actual and estimated efforts and size for the respective projects that have been identified in order to draw analogies. The mean magnitude of the relative error in the effort estimation of the

projects is used to build contingency and to adjust the UCP for the purpose of the project proposal.

(c) Use the effort to arrive at the duration of the project, broken down by phases. The number of resources to be deployed per phase is estimate.

(d) Calculating the project cost or price of the bid on the basis of the costs associated with the resource, travel, training and other indirect costs if any.

Example

In the rest of this section the implementation of the above three steps is demonstrated using a hypothetical example

Step 1: Business domain experts take the inputs from the RFP and use cases as available in the reference catalogue to produce a "minimal use cases set" as necessary for the software to be developed. The team brainstorms, researches or uses industry models to find additional use cases to define the unstated requirements.

For example, a client forwards a one-page RFP asking for the cost of building and deploying a system that can help the CIO to centralise the information of the company's IT assets, to plan initiatives and to align them with the business strategy. This project, referred to as the "IT Applications Management Dashboard (IAMD)", is to be estimated in the absence of detailed requirements.

The business domain experts may refer to the use case catalogue to come up with the possible use cases or business scenarios to be supported. Similar user cases might have been captured and coded earlier in previous projects and could be reused.

The use case catalogue should categorise the use cases employing requirement patterns. The requirement pattern could (for beginners) be very generic, for example "capture actor details", "capture program details", "alerts", "workflow", "report generation", "context-sensitive help" etc. Organisations with higher maturity levels may be expected to have specific patterns depending upon the business domain; for example, "actor details" could be further categorised as "broker details" where matters involve the financial services business domain.

Let's assume that on completion of Step 1 we have the minimal set of use cases shown in Table 7.1.

Step 2: Application domain experts extend the use case set, employing information about the past projects available in the database. The team draw analogies on the basis of various dimensions of software development and come up with the ranked list of projects that are nearest to the project to be developed. The project database is classified on multiple dimensions and each project in the database is ranked on the basis of its value for these dimensions. Figure 7.6 shows an example of classifying the project database (using 13 dimensions grouped under 3 aspects, i.e. functional, technological and process or environmental).

Table 7.1 Set of minimal use cases.

No.	*Description*	*Status*
1	Add/update scorecard	New
2	Add/update initiative anchors' details	To be reused
3	Add/update initiative or proposal details	To be reused
4	Add/update prioritization parameters	New
5	Add/set alerts for variances	To be reused
6	Generate alert-based communication workflow	To be reused
7	Facilitate prioritization of initiatives	New
8	Add/capture progress report on each initiative	To be reused
9	Facilitate project planning for each initiative	New
10	Roll up tracking metrics to CIO level	New

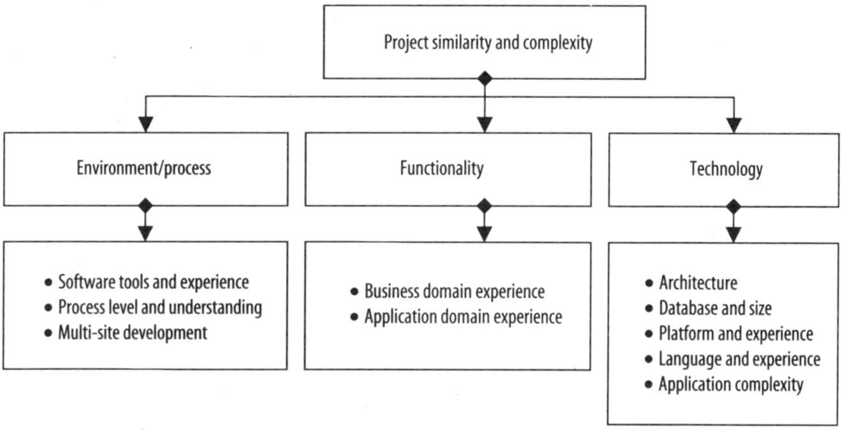

Figure 7.6 Historical data and classification schema.

The organisation assigns relative weights to these dimensions according to complexity (i.e. risk), on the basis of in-house competencies (e.g. skilled professionals), maturity level (e.g. the scope of the reusability of software assets) and experience (e.g. the number of projects executed in the past) etc. AHP can be used to rank these dimensions. All the projects in the database thus have a rank. The information related to complexity, process, technology and functionality is available for conducting searches and drawing analogies.

The relative weights may be revised as and when there is a change in the maturity level (example: the inclusion of tools or best practices), experience (example: new projects) and in-house competencies (example: new skills and/or number of resources) etc.

For example, an organisation which is well equipped with technical and process-related skills would place greater emphasis on functionality. Table 7.2 provides the relative weights of dimensions. The higher the weight, the more critical is the dimension for the organisation and success of the project. Each of these dimensions can be

Table 7.2 Dimensions and their relative importance in classifying projects.

Dimension		*Weight*	*Sub-dimension*		*Weight*
Process	F1	0.088	Software tools and experience	F1.1	0.443
			Process understanding	F1.2	0.387
			Multi-site development	F1.3	0.170
Functionality	F2	0.669	Business domain knowledge	F2.1	0.250
			Application domain knowledge	F2.2	0.750
Technology	F3	0.243	Architecture	F3.1	0.148
			Database and experience	F3.2	0.110
			Platform and experience	F3.3	0.080
			Language and experience	F3.4	0.331
			Application complexity	F3.5	0.331

further broken down into the sub-dimensions that constitute them. The same philosophy is used in giving relative weights to the sub-dimensions.

The "IAMD" project team can select the most appropriate value (defined as "class" in Table 7.3) for the sub-dimensions. These values also form the search criteria to identify similar projects executed in the past. The similarity is in terms of the risks and the complexity and also in terms of the following:

1. The environment in which the project is being developed, i.e. the tools, the process and its maturity level, as well as the number of coordinated locations etc.
2. The functionality, i.e. the business domain (examples: financial services, manufacturing etc.) and application domain (examples: customer relationship management, enterprise resource planning, business intelligence etc.).
3. The technology, i.e. the application architecture, database, programming language and complexity of logic (examples: simple, procedural, algorithmic, device-dependant etc.).

Each project in the database possesses a cumulative rank for overall similarity in terms of class and complexity; it may be defined in terms of a cumulative similarity index, calculated as follows:

$$\textit{Cumulative Similarity Index}(CSI) = \sum (F_i \times F_{i,j,k}) \times (l)$$

where F_i represents dimension, $F_{i,j}$ represents sub-dimension under F_i, $F_{i,j,k}$ represents class of sub-dimension and l represents the value as 1 or 0 based on whether the project falls in the class or not.

For example, the current IAMD project (based on the characteristics defined in Table 7.4) has a CSI of 0.2245:

CSI(IAMD) $= 0.088 \times 0.443 \times 0.105 + 0.088 \times 0.387 \times 0.152 + 0.088 \times 0.170 \times 0.163 + 0.669 \times 0.250 \times 0.285 + 0.669 \times 0.750 \times 0.188 + 0.243 \times 0.148 \times 0.731 + 0.243 \times 0.110 \times 0.072 + 0.243 \times 0.080 \times 0.257 + 0.243 \times 0.331 \times 0.188 + 0.243 \times 0.331 \times 0.279 = 0.2245$

Table 7.3 Sub-dimensions, values and relative weights.

Sub-dimension	*Class*	*Weight*	
F1: Environment/process dimension			
Software tools and experience	F1.1.1	<1 year	0.105
	F1.1.2	1–3 years	0.258
	F1.1.3	>3 years	0.637
Process understanding	F1.2.1	Level 1	0.052
	F1.2.2	Level 2	0.089
	F1.2.3	Level 3	0.152
	F1.2.4	Level 4	0.262
	F1.2.5	Level 5	0.445
Multi-site development	F1.3.1	Global	0.540
	F1.3.2	Country-localised	0.297
	F1.3.3	City-localised	0.163
F2: Functionality dimension			
Business domain	F2.1.1	Thorough understanding	0.637
	F2.1.2	Reasonable understanding	0.285
	F2.1.3	No understanding	0.105
Application domain	F2.2.1	Thorough understanding	0.731
	F2.2.2	Reasonable understanding	0.188
	F2.2.3	No understanding	0.081
F3: Technology dimension			
Architecture	F3.1.1	Mainframe	0.081
	F3.1.2	Client-server	0.188
	F3.1.3	Multi-tier	0.731
Database and size	F3.2.1	Small-medium	0.072
	F3.2.2	Large	0.279
	F3.2.3	Very large	0.649
Platform and experience	F3.3.1	<1 years	0.152
	F3.3.2	1–3 years	0.257
	F3.3.3	>3 years	0.591
Language and experience	F3.4.1	<1 year	0.081
	F3.4.2	1–3 years	0.188
	F3.4.3	>3 years	0.731
Application complexity	F3.5.1	Simple	0.072
	F3.5.2	Moderate	0.279
	F3.5.3	Complex	0.649

Table 7.4 Characteristics of IAMD.

No.	*Sub dimensions*	*Values*
1	Software tools and experience	<1 year
2	Process understanding	Level 3
3	Multi-site development	City-localised
4	Business domain	• Reasonable understanding • Retail
5	Application domain	• Reasonable understanding • Business intelligence
6	Architecture	Multi-tier
7	Database and size	• Small-medium • Oracle 8i
8	Platform and experience	• 1–3 years • Windows 2000
9	Language and experience	• 1–3 Years • ASP • HTML
10	Application complexity	Moderate

Table 7.5 Result set.

Project metric		*Project*			
		A	*B*	*C*	*D*
Size (UCP)	Estimated	480	600	480	640
	Actual	525	560	420	630
Effort (person-days)	Estimated	360	450	360	480
	Actual	450	480	360	540

Table 7.6 Business requirements/uses cases and complexity.

No.	*Description*	*Complexity*
1	Add/update scorecard	Complex
2	Add/update initiative anchors' details	Medium
3	Add/update initiative or proposal details	Medium
4	Add/update prioritization parameters	Complex
5	Add/set alerts for variances	Complex
6	Generate alert-based communication workflow	Complex
7	Facilitate prioritization of initiatives	Complex
8	Add/capture progress report on each initiative	Medium
9	Facilitate project planning for each initiative	Complex
10	Roll up tracking metrics to CIO level	Medium
11	Maintain project tracking metrics per initiative	Medium
12	Add/update prioritization rank of initiatives	Medium
13	View/update reallocate initiatives	Simple
14	View/update budget allocation and cash-flow planning	Simple
15	View/update organization-wide audits and reviews	Simple
16	Add/help PMO to identify and announce risks and rewards	Complex
17	Generate reports at each user level	Medium
18	Integrate with business intelligence system for corporate inputs	Complex
19	Facilitate/update project tracking information per initiative	Complex
20	Maintain history of events including reviews, audits, creation, realization etc.	Complex
21	Automate communication such as distribution of reports, participation requests etc. through workflow	Complex
22	Integrate with market intelligence system for surveys, reports, benchmarks and information updates of competitors' moves	Complex
23	Facilitate joint decisions, i.e. multi-party decisions or surveys using the AHP framework	Complex
24	Navigate using drill-down views	Complex

One can also compare the projects using an individual similarity index (ISI), i.e. at either individual dimension or sub-dimension level.

The characteristics of IAMD can be used as search criteria (Table 7.4) and to draw analogies with the projects executed in past. Table 7.5 shows the representative set of projects with the same CSI or a CSI very close to that of IAMD, i.e. 0.2245.

Application domain experts may brainstorm and use inputs from Projects A, B, C and D to come up with additional use cases that fit the project profile (IAMD).

The minimal set of use cases as outlined by the business domain experts earlier in Step 1 can now be updated on the basis of past projects. The final and complete set of use cases can be finally collated and sized as shown in Table 7.6.

Table 7.7 Actors and complexity.

No.	*Actors*	*Complexity*	*Weight*
1	CIO – GUI-based interface	Complex	3
2	Senior IT management – GUI-based interface	Complex	3
3	Program managers/initiative anchors – GUI-based interface	Complex	3
4	Email system – API-based interface	Simple	1
5	Business intelligence system – interface through TCP/IP	Medium	2
6	Market intelligence system – interface based through TCP/IP	Medium	2

Table 7.8 Estimated build effort.

Use case type	*Weights*	*Number of units*	*Total size**
Simple	5	3	15
Medium	10	7	70
Complex	15	14	210
	Total (person-days)	295	

*Per use case type.

One can estimate the size based on the relative complexity of the 24 use cases as listed in Table 7.6, and the 6 actors involved in the usage of the system in Table 7.7.

The estimate for the size based on the use case is captured in Table 7.8.

The software size estimate in UUCP, based on the number and relative complexity of the use cases as given in Table 7.8 and the relative complexity of actors as given in Table 7.7, is computed as follows:

$$UUCP = 14(= 3 \times 3 + 1 \times 1 + 2 \times 2) + 295$$
$$(= 5 \times 3 + 10 \times 7 + 15 \times 14) = 309$$

If a number of application domain experts are forming an opinion about the size of the project, then there should be a way to arrive at a single estimate at the end. This requires a systematic and logical approach to the ranking of the sources of the estimate, i.e. the experts. Thus apart from classifying projects, there is a need for suitable criteria to rank the estimates provided by the experts (see Figure 7.7).

Table 7.9 provides the estimates arrived by the experts, based on their rating for the complexity of the use cases and the actors to arrive at an estimate.

Using the expert opinions and the relative rankings of the experts, one can arrive at a UUCP (adjusted for expert opinion) as follows:

$$UUCP = 309 \times 0.281 + 334 \times 0.319 + 299 \times 0.400 = 313$$

Step 3: In Step 3, the technology expert uses the project (result set) information and the information in RFP to assign values to the TCF and EF.

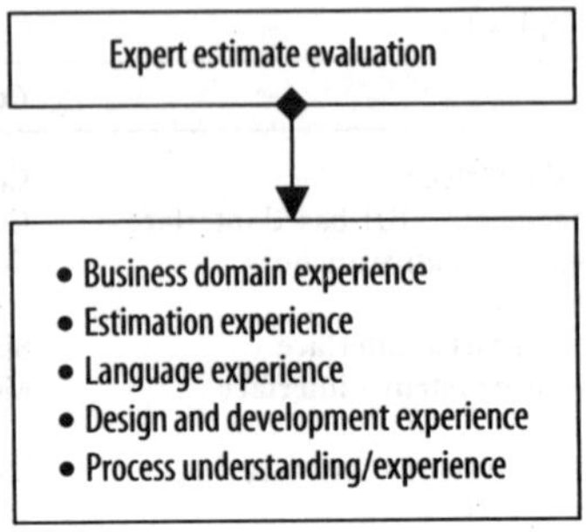

Figure 7.7 Relative ranking of expert estimates.

Table 7.9 Relative ranking of experts.

Expert	*Estimated effort*	*NEV*
Application domain expert 1	309	0.281
Application domain expert 2	334	0.319
Application domain expert 3	299	0.400

Assume that for IAMD, the technology expert may choose the values for degree of relevance of 13 technical complexity factors as $T_1 = 4$, $T_2 = 3$, $T_3 = 5$, $T_4 = 3$, $T_5 = 5$, $T_6 = 1$, $T_7 = 1$, $T_8 = 2$, $T_9 = 3$, $T_{10} = 2$, $T_{11} = 3$, $T_{12} = 0$ and $T_{13} = 1$, and 8 environmental factors as $E_1 = 2, E_2 = 3, E_3 = 3, E_4 = 3, E_5 = 4, E_6 = 3, E_7 = 4$ and $E_8 = 3$.

The size for IAMD in UCP is

$$\begin{aligned} UCP &= 313 + [0.6 + (0.01) = (4 = 2 + 3 = 2 + 5 = 1 + 3 \\ &= 1 + 5 = 1 + 1 = 0.5 + 1 = 0.5 + 2 = 2 + 3 = 1 + 2 \\ &= 1 + 3 = 1 + 0 = 1 + 1 = 1)] + \{1.4 + (-0.03) \\ &= [2 = 1.5 + 3 = 0.5 + 3 = 1 + 3 = 0.5 + 4 = 1 + 3 \\ &= 2 + 4 = (-1) + 3 = 2]\} = 314.78 \end{aligned}$$

From the proposal perspective, the size estimate can be adjusted for the estimation accuracy. The estimation accuracy is a comparison of the estimated and the actual size measured using formulas such as:

1. Relative error (RE): $(E_{est} - E_{act})/E_{act}$.
2. Magnitude of relative error (MRE): $|RE|$.
3. Mean magnitude of relative error (MMRE): $(100/n) \times \Sigma MREI$, where $i = 0 \ldots n$ and n denotes the number of projects.

Table 7.10 provides the estimation accuracy of the projects of the result set.

The size in UCP, after adjusting for the estimation accuracy, is

$$\begin{aligned} &UCP \text{ (after factoring estimation accuracy)} \\ &= 314.78 \times 1.0792 = 339.71 \end{aligned}$$

Table 7.10 Estimation accuracy of the projects.

Project metric		*Project*			
		A	*B*	*C*	*D*
Size (UCP)	Estimated	480	600	480	640
	Actual	525	560	420	630
Effort (person-days)	Estimated	360	450	360	480
	Actual	450	480	360	540
Magnitude of size relative error		0.086	0.071	0.14	0.02
Magnitude of effort relative error		0.20	0.06	0	0.11
Mean magnitude of size relative error			7.92		
Mean magnitude of effort relative error			9.25		

Table 7.11 Phasewise effort estimate and deployment of resources.

Activity	*Estimated effort (person-days)*			
	% of total	*Effort*	*Resources deployed*	*Avg. cost ($) per resource per day*
Requirements	10	28.87	4	120
Design	12	34.65	4	100
Build	29	83.74	8	95
Integration testing	7	20.21	2	95
Regression testing	2	5.80	2	95
Acceptance testing	6	17.32	2	95
Project management	15	43.31	2	140
Configuration management	3	8.67	2	75
Training	10	28.85	2	75
Others	6	17.32	2	75
Estimated effort (%)	100	288.74		

Resource cost ($) = (Effort (Average resource cost (Number of days).

Using the baseline effort of 0.85 person-days (assuming 8 hours per day) per UCP, the effort for IAMD can be estimated to be 288.75 person-days.

Table 7.11 provides the estimated breakup of the effort in terms of the phases of the project and also the estimated number of resources.

From Table 7.11, the cost of resources deployed can be calculated as follows:

$$\text{Resource cost (\$)} = 29\ 178$$

The proposed price for the project should include other costs such as training and travel. In case the project needs extra hardware or software to be procured, the same should be factored into the cost. The price may be inflated with buffers for contingency and negotiation, prior to adding the profit margin.

Assuming the values for training as $2000, travel as $5000, a contingency buffer of 10%, a negotiation buffer of 10% and a profit margin of 35%, the proposed price will be

$$\text{Bid/proposed price (\$)} = 58\ 608$$

7.4 Conclusion

The approach described above highlights some of the best practices to ensure that the application development costs are calculated with great accuracy even in the absence of sufficient data at the proposal submission stage. The best practices are as follows:

1. Building a repository of reusable IT assets; for size and effort estimation purposes, the repository implies cataloguing use cases or function point models based on the requirement patterns.
2. Ranking projects based on the risks and complexities; use of the AHP to evaluate the projects can help the organisation compare the use cases across the business and application domains, by ranking the domains in terms of their relative similarity between them.
3. Building a project database and storing information on the project across its life cycle.
4. Enriching the reference catalogue by involving the domain experts from the very beginning of the proposal preparation process. This constitutes an addition of value to the organisation even if the proposal fails to get converted.
5. Ranking the experts in order to encourage and systematically increase their involvement.

A reference model or reference use case catalogue may seem to be an overhead, but such an asset can gradually improve the accuracy of estimation accuracy in cases where only limited data are available. Moreover, it helps the development team to understand the functionality well ahead of the software development. The best practice of cross-referencing these reference models and/or use cases to their implementation artefacts helps to reuse the assets developed in the past. A good framework for classifying and storing the project data is just as essential as the data themselves. The framework should help to categorise the projects according to a number of dimensions.

References

1. Boehm BW (1984) Software engineering economics. IEEE Transactions on Software Engineering SE-10(1).
2. Graham C, Low D, Ross J (1990) Function points in the estimation and evaluation of the software process. IEEE Transactions on Software Engineering 16(1).
3. Function Point Counting Practices Manual (1999) Release 4.1.1, IFPUG, January.
4. Ferrentino AB, Making software development estimates "good". Tutorial: Software Engineering Project Management by Richard H. Thayer, IEEE Computer Society Press.
5. Boehm BW (1984) Software engineering economics. IEEE Transactions on Software Engineering SE-10(1).
6. Tridas Mukhopadhyay and Sunder Kekre (1992) Software effort models for early estimation of process control applications. IEEE Transactions on Software Engineering 18(1).
7. Boehm BW (1984) Software engineering economics. IEEE Transactions on Software Engineering SE-10(1).

Appendix 7.1 Ranking Projects and Experts Using the AHP

Application domain experts can rank the project relatively on multiple dimensions to search the project database for similar set of applications developed earlier.

I Project Similarity and Complexity Factors

	Functionality	*Technology*	*Process*	*NEV*
Functionality	1	3	7	0.669
Technology	1/3	1	3	0.243
Process	1/7	1/3	1	0.088

II Functionality

	Business domain	*Application domain*	*NEV*
Business domain	1	1/3	0.250
Application domain	3	1	0.750

III Technology

	A	*D*	*L*	*P*	*C*	*NEV*
A	1	2	1/3	2	1/3	0.148
D	1/2	1	1/3	2	1/3	0.111
L	3	3	1	3	1	0.331
P	1/3	1/2	1/3	1	1/3	0.080
C	3	3	1	3	1	0.331

A, architecture; D, database and size; L, language and experience; P, platform and experience; C, application complexity.

IV Process

	S	*P*	*M*	*NEV*
S	1	1	3	0.443
P	1	1	2	0.387
M	1/3	1/2	1	0.169

S, software process tools and understanding; P, process understanding; M, multi-site development.

V Business domain

	T	*R*	*N*	*NEV*
T	1	3	5	0.105
R	1/3	1	3	0.258
N	1/5	3	1	0.637

T, thorough understanding; R, reasonable understanding; N, no understanding.

VI Application domain

	T	*R*	*N*	*NEV*
T	1	5	7	0.731
R	1/5	1	3	0.188
N	1/7	1/3	1	0.081

T, thorough understanding; R, reasonable understanding; N, no understanding.

VII Architecture

	Mainframe	*Client server*	*Multi-tier*	*NEV*
Mainframe	1	1/3	1/7	0.081
Client Server	3	1	1/5	0.188
Multi-tier	7	5	1	0.731

VIII Database/size

	Small-medium	*Large*	*Very large*	*NEV*
Small-medium	1	1/5	1/7	0.072
Large	5	1	1/3	0.279
Very large	7	3	1	0.649

IX Application complexity

	Simple	*Moderate*	*Complex*	*NEV*
Simple	1	1/5	1/7	0.072
Moderate	5	1	1/3	0.279
Complex	7	3	1	0.649

X Programming experience

	<1 year	*1–3 years*	*>3 years*	*NEV*
<1 year	1	1/3	1/7	0.081
1–3 years	3	1	1/5	0.188
>3 years	7	5	1	0.731

XI Application experience

	<1 year	*1-3 years*	*<3 years*	*NEV*
<1 year	1	1/3	1/7	0.081
1–3 years	3	1	1/5	0.188
–3 years	7	5	1	0.731

XII Platform experience

	<1 year	*1–3 years*	*>3 years*	*NEV*
<1 year	1	1/2	1/3	0.152
1–3 years	2	1	1/2	0.257
>3 years	3	2	1	0.591

XIII Software tool and experience

	<1 year	*1–3 years*	*>3 years*	*NEV*
<1 year	1	1/3	1/5	0.105
1–3 years	3	1	1/3	0.258
>3 years	5	3	1	0.637

XIV Process maturity

	Level 1	*Level 2*	*Level 3*	*Level 4*	*Level 5*	*NEV*
Level 1	1	1/2	1/3	1/5	1/7	0.052
Level 2	2	1	1/2	1/3	1/5	0.089
Level 3	3	2	1	1/2	1/3	0.152
Level 4	5	3	2	1	1/2	0.262
Level 5	7	5	3	2	1	0.445

XV Multi-site development

	I	*N*	*L*	*NEV*
I	1	2	3	0.540
N	1/2	1	2	0.297
L	1/3	1/2	1	0.163

I, international/ multiple nations; N, national and multiple cities; L, localised within city;

Ranking Experts

I Expert estimate ranking attributes

Level 1	*B*	*E*	*L*	*D*	*P*	*NEV*
B	1	2	1/3	2	1/3	0.148
E	1/2	1	1/3	2	1/3	0.111
L	3	3	1	3	1	0.331
D	1/3	1/2	1/3	1	1/3	0.080
P	3	3	1	3	1	0.331

B, business domain experience; E, estimation experience; L, language and experience; D, design and development experience; P, process understanding/ experience.

II Ranking experts on business domain experience

Level 2	*Expert 1*	*Expert 2*	*Expert 3*	*NEV*
Expert 1	1	1	3	0.443
Expert 2	1	1	2	0.387
Expert 3	1/3	1/2	1	0.169

III Ranking experts on estimation experience (accuracy in past)

Level 2	*Expert 1*	*Expert 2*	*Expert 3*	*NEV*
Expert 1	1	1/3	1/5	0.105
Expert 2	3	1	1/3	0.258
Expert 3	5	3	1	0.637

IV Ranking experts on language and experience

Level 2	*Expert 1*	*Expert 2*	*Expert 3*	*NEV*
Expert 1	1	1/2	1/3	0.152
Expert 2	2	1	1/2	0.257
Expert 3	3	2	1	0.591

V(a) Ranking experts on design and development experience

Level 2	*Expert 1*	*Expert 2*	*Expert 3*	*NEV*
Expert 1	1	1/3	1/5	0.105
Expert 2	3	1	1/3	0.258
Expert 3	5	3	1	0.637

V(b) Ranking experts on process understanding and experience

Level 2	*Expert 1*	*Expert 2*	*Expert 3*	*NEV*
Expert 1	1	1	3	0.443
Expert 2	1	1	2	0.387
Expert 3	1/3	1/2	1	0.169

VI Overall ranking of experts

Level 2	*Rank*
Expert 1	$0.148 \times 0.443 + 0.111 \times 0.105 + 0.331 \times 0.152 + 0.080 \times 0.105 + 0.331 \times 0.443 = 0.281$
Expert 2	$0.148 \times 0.387 + 0.111 \times 0.258 + 0.331 \times 0.257 + 0.080 \times 0.258 + 0.331 \times 0.387 = 0.319$
Expert 3	$0.148 \times 0.169 + 0.111 \times 0.637 + 0.331 \times 0.591 + 0.080 \times 0.637 + 0.331 \times 0.169 = 0.400$

PART 3

Strategic Decision-Making in Defense and Governance

8 Prioritising National Security Requirements

8.1 What Is National Security?

Despite the centuries of civilisation, the human race is still unable to avoid war. Wars will continue while human beings exist. Therefore, the study of ways of avoiding wars, of prosecuting them if they are imposed, of sustaining and supporting them, is an essential activity of a modern nation. The knowledge-based world of today will influence the future way the security of a nation is threatened and that threat responded to. Future trends in a nation's security will be greatly affected by the fundamental changes taking place in the global geo-strategic environment and the increasing role of political, cultural and socio-economic factors in the overall security equations. However, technology will have sweeping influences on future security scenarios. Therefore, to understand a country's future security requirements it is necessary not only to understand the unpredictable, uncertain and unknown domains into which our world is rapidly moving, but also to assess the revolutionary ways technological developments will influence future security needs and scenarios.

Owing to increasing globalisation a nation in today's world can not remain unaffected by events on other parts of the earth. The security environment of a country is impacted by geopolitical factors, social structures and economic conditions. However, technological factors are increasingly having a major influence on any country's national security. How the power structure of the world changes, and in which dimensions it is manifested, may be the most important factor in the future security of a nation. During strategic national security planning it is imperative to make a comprehensive assessment of both the prevailing situation and the future contingencies that it may be necessary to cope with. This assessment requires a systematic analysis of trends, events and factors that may impact a nation's security in both the medium and long terms. To carry out such an analysis it is relevant to understand trends in the many dimensions in which our world is moving that may have critical implications for national security. It is necessary to have a detailed map of the future world structure.

8.2 The Future World Structure

Scientists, strategists, politicians and managers are forced to look into the future to gain the understanding they need in order to take strategic decisions that will impact favourably on their country. It is clear that the future is not knowable, but

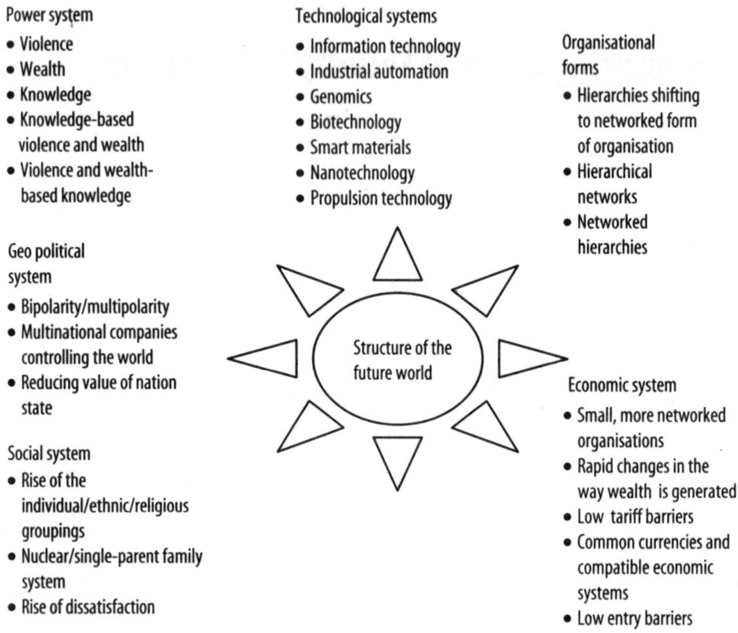

Figure 8.1 Structure of the future world.

"It is better to have a general and incomplete map (of the future) subject to revision and correction, than to have no map at all" [1]. Here we will try to build up a map of the future world that is now approaching us. Unless such an exercise is performed the ways to deal with uncertain and unknown future threats to national security cannot be planned or calculated.

The pace of change today is so intense that any prediction of the future world structure must be taken with a pinch of salt. However, with imagination and an analysis of current situations one can decipher the general trends. The fundamental prerequisite is the identification of the basic parameters that should be studied for the prediction of future world structures. The underlying structures that will shape the future world are power systems, technological systems, geopolitical systems, social systems, evolving organisational forms and commercial/economic systems (see Figure 8.1). It must be emphasised that these are not independent systems but different ways of looking at general world trends. There are links between these systems, which affect each of them in many dimensions. We will study each one of them as an independent system for the purpose of identifying its influence on national security.

8.2.1 Power Systems

One most important point to be understood is that the universe is not at all 'deterministic', as was believed by Western science for past so many centuries. The order, predictability and estimation of cause effect relationships between various events and conditions in this world is not possible. However, the universe is not completely "random". Time predictions on the basis of past events are generally possible. The fact that the universe is neither totally random nor totally deterministic creates the option of using power to change the events that occur in it. This ability to change the events gives a necessary impetus to individuals, nations and group of nations to seek more power. Power therefore combines both chaos and order. A society should strive to have a judicious mix of these two diverse phenomena. The individual or the nation that controls these two opposites so as to maximise the benefits will gain more power than one that is tilted towards either of these distinct poles.

The most important, current and flexible way of gaining power is through knowledge. Violence and wealth as tools of exercising control have already proved their limitations. With enough knowledge it is possible to control even violence and wealth. The most important factor, which is already characterising the power systems of the world, is the ability to gain and apply knowledge. This shift in the power system of the world from violence and wealth to knowledge can be easily seen in many walks of life. [2] The trend is also visible in warfare, which by its very nature is a violent activity and concerns the use of violence as a means of applying power. Yet in this predominantly violent activity, appropriate and timely knowledge of the enemy has become an almost decisive factor. The interplay of knowledge, violence and wealth will create multiple dimensions in which power can be gained and exercised. This may be achieved through knowledge-based violence or knowledge-based wealth. On the other hand, knowledge relevant for violence and knowledge relevant for wealth creation may provide power to individuals, groups or nations. Therefore in future the power system will be based more on knowledge than on mere wealth and violence, as was the case for many centuries past [2].

8.2.2 Technological Systems

The shift in the base of power systems from wealth and violence to knowledge has come about owing to revolutionary changes made in information technology (IT). The world is moving at a rapid pace. Speed has become the buzzword in every human endeavour. This creates a need not only for the speedy flow of information but also for its availability to people at all times. In all walks of life the ability to acquire information makes a major difference in achieving success in today's world of cut-throat competition, and it can now be done at great speed thanks to the massive advances in IT.

The performance of computer microprocessors is doubling each year, thereby providing powerful applications on inexpensive machines. This, coupled with the growing maturity in the field of parallel processing, is providing powerful computing machines. Similarly, computer peripherals are increasing in speed and performance. The future computing system will be based on powerful processors, capable of handling terabytes of data per second coupled with peripherals such as disk arrays, networks and high-performance graphic systems.

Putting semantics to symbolic information leads to knowledge, which is defined as the ability to derive inference from information and data. Knowledge representation, its efficient storage, retrieval and usage, has led to what is known as knowledge processing. Together with high-speed computing machines, knowledge processing has eased the job of decision-makers in various management and professional fields. This has led to the development of the expert system (ES) – the knowledge base of a particular application domain together with an inference engine guided by well-established production rules and a smart graphical user interface (GUI).

The progression of networking and communication, from the individual stand-alone computing system to the resource-sharing machine (the networking system with interconnections), to the internet, has revolutionised the world. By developing a set of rules (protocols), for how computers (hosts) communicate over a network and further rules for connecting one type of network to another, it is possible to build interconnected sets of networks in which any computer can communicate with any other computer.

The dramatic increase in computing performance has encouraged researchers to enhance internetworking services. The powerful computers of today can support interactive applications such as simulations, multimedia conferencing and virtual reality. The requirement of future internetworking protocols is to provide security and performance guarantees to applications. The information superhighways have created a remarkable capability to gain information in near real time. This capability is increasing day by day. The future of computer networking lies in an intelligent network, which is aware of what it is doing. A network that knows what it is sending and continuously monitors its performance is not beyond the bounds of today's technology and is the possible future. When these networks combine with the advances made in the security and coding of messages their utility increases many times over.

Another technological area that is influencing the world is that of precision guidance, automation and robotics. Advances in this field provide increasing precision in controlling the movement and speed of physical objects. The increasing precision is providing pinpoint accuracy with which physical objects can reach a particular point. The underlying trend in the great advances achieved in IT and precision guidance is the increasing levels of miniaturisation. These technologies are combining to give an enormous capability in surveillance and information gathering. Another area of great strides in performance is propulsion technology. Propulsion systems are creating rapid transport capabilities in the air, over land and sea, and under the sea. These have major implications for the rapid mass transport of men and materials.

The twenty-first century has been called the century of biology. This is because of the immense potential of biotechnology that is evident. The "revolution of living things" will lead to increased quantity and quality of human life [3]. Better disease control, custom drugs, intelligent drug-delivery systems, gene therapy, age mitigation and reversal, memory drugs, bionic implants etc. are all now considered likely to be the normal, day-to-day applications of the future world. The revolution in materials, devices and manufacturing will produce smaller, smarter and more flexible systems. These will impact manufacturing, logistics and personal lifestyles. Smart wearable materials, agile manufacturing, nanofabricated semiconductors and integrated microsystems are some of the realties of the future world. These technological revolutions will impact the world at large in many dimensions and in different ways and to different extents on different parts of the globe. The RAND report details the meta-trends and implications as follows. The accelerating pace of technological change combined with creative destruction will increase the importance of continuous learning, the nature of technology will become increasingly multidisciplinary, and the world will be driven by the competition for leadership in technological development, by continued globalisation, and by the latent lateral penetration of technology into entirely new areas. However these trends will create major tensions owing to class disparities, reduced privacy, and the perceived threats which technological changes will bring to existing cultural systems in various parts of the world.

Although technological revolutions are influencing the way of the future world, the geopolitical realities will nevertheless not go away in a hurry. The cold war of the last century has left its legacy to the present millennium in the form of terrorism and fundamentalism. The technological advances will help non-state actors, e.g. terrorist groups, to create innovative acts of terror – which may employ very simple disruptive mechanisms or highly sophisticated large-scale means of destroying life and property. These will create asymmetrical conflicts where the state actors are technologically and operationally equipped to fight more traditional wars while the new technologies create non-conventional capabilities with the non-state actors.

8.2.3 The Geopolitical System

With the demise of the USSR, the bipolar stability of the cold war years vanished. The new world found itself staring at a period of possible freedom and peace. What the Soviet Union and the United States of America, as the two counterbalancing superpowers, achieved was a stable world in which the number of conflicts may have been great but large-scale wars were few in number. The Iraqi annexation of Kuwait in 1990 however showed the likely dangers of a world where the balancing forces of superpowers have gone. The 1991 Gulf War proved to be a turning point in the history of warfare. It combined the doctrine of the attritional warfare of so many centuries in the past with the new military doctrine of making surgical strikes against the enemy with high-technology weapons, sensors and missiles. This was the beginning of what has been called the latest revolution in military affairs (RMA).

The Gulf War showed to the world the awesome military power of the sole superpower. The war proved the all-pervading capability of the USA. However, it showed to the other world powers that their military structures and organisations based on old doctrines cannot compete with the new technology of warfare. This has led to increasing high-technology weapon acquisition by the other power centres of the world. Given this development we have to work out what sort of political system we will be looking at in future.

Most experts believe we are moving towards a multipolar world. The possible cooperation between China and Russia as a bloc against the USA will lead to the strengthening of China. After twenty years or so, China may become a power centre to challenge the USA – a pole (power centre) in a pentapolar world composed of the USA, the European Community, Japan, China and Russia. There may be players other than these five, but it is likely that world will have several power centres.

Also, the world of large, monolithic and self-sufficient nations or organisations is withering away. Every nation needs others to sustain it and enable it to grow. Only through mutual growth based on sound symbiotic relationships can a nation develop. The world is beginning to free itself from the fissures of national boundaries. It is in regional groupings based on geographic locations, economic requirements or social obligations that nations are beginning to find more benefits. The future poles or power centres of the world may be groups of nations organised as economic, military or geographic confederations against other such groups or against single large and powerful nations such as the USA or China.

The disintegration of powerful empires into smaller, more manageable units is also visible in the evolving social systems of the world.

8.2.4 The Social System

In the Western world large families have been replaced by individual nuclear families. The same trend is visible in Asian families, which were bound together initially but are now finding themselves led into smaller more manageable units. The trend is similar to the breakdown of USSR: the world is shifting towards individual units. The importance each individual seeks for themself is so overriding that the individual feels compelled to leave the joint family system. This need for importance is not necessarily a search for an identity for the individual but may represent a quest for economic independence. Asian families are closely bound together for cultural and religious reasons, yet in future more and more families will be either nuclear or single-parent families. The trend can be seen in large cities. Exodus from the villages to the cities is increasing owing to poor returns from agriculture and village-based small industries. This will lead to the overcrowding of big cities and to many more cities growing up within the next two decades.

Another trend that can be discerned is the rise of the middle class as a powerful influence on the economic, political and military structures of every nation in the modern world. This has led to growing markets for multinational companies and a rising consumerism all over the world. For economic reasons the multinationals may try to influence the geopolitical situations and structures of the world. A very important and powerful pole or power centre of a future multipolar world may be a consortium of multinational companies whose products involve almost the whole of the world and whose interests lie in maintaining a particular status in a particular environment. This may lead to possible conflicts with other such groups or even with nation states.

Communalism and a return to religious uprisings are also possible in future. Any individual, whatever the lure of materialistic wealth, still feels a sense of longing for his roots based on the prevalent religious, cultural and social structures. The return to religious groups and the associated rise in fundamentalism are unique phenomena for Third-World countries. This trend, fuelled by growing populations and the associated frustrations, may lead to an alarming increase in terrorism and insurgency in the world. We may face an increasing number of low-intensity conflicts (LIC) in future. An LIC is defined by [4] as an armed conflict for political purposes short of combat between regularly organised forces. It is distinct from mid-intensity conflict (MIC), which is armed combat between regularly organised military forces. High-intensity conflict (HIC) is armed conflict involving the use of weapons of mass destruction (WMD).

8.2.5 Evolving Organisational Forms

The information revolution favours the rise of the network form of organisation [5]. The network is the most modern form of organisation that human civilisation has seen, the successor to tribes, hierarchies and markets. This new form of organisation can compete with the hierarchical structure. Since most nations have governing structures which are hierarchical in nature, the security of a nation state will be threatened if the nation is forced to face a network form of organisation, e.g. the terrorist groups forming a network of small units that swarm to the point of attack and disperse immediately. The forms of organisation will have a telling effect on various structures in the world scenario. It is likely that networks and hierarchies will create hybrid structures as well, e.g., networked hierarchies and hierarchical networks.

8.2.6 Economic System

The new economic system of the world is based on "softnomics" – media, communications, and computers, i.e. software and services. It is in stark contrast to the old system based on hardware and manufacturing. Today's companies are striving to cope up with an info-technological revolution, the restructuring of markets and the powerful influence of marketing on business returns. The new economy, characterised by takeovers, financial liberalisation and reshuffling, is totally dependent upon the timely, pinpointed availability of information. Knowledge has become the key weapon in the power struggles that accompany the new economic structure of the world.

With the demise of communism in Russia and the opening of the economies of China and India to provide large consumer markets, the business world is going through a period of great expansion. The multinational companies that can cope up with the diverse cultures and problems of operating in numerous countries will survive and progress, but others will have to reorient their strategies to operate within the constraints of the uncertainties and instabilities associated with multiple markets.

No single company will be able to deal with the diverse problems indicating the demise of large multinational companies in future. However, this may lead to a consortium or a loosely coupled network of companies joining together in every part of the world for the sake of mutual beneficial economic and commercial gain. This could be the scenario within the next two decades, if the top CEOs realise that conflicts with their peer companies will lead to mutual destruction. Such a trend will be similar to the superpower arms talk of the cold war period, when the realisation of the horrifying potential consequences of an inadvertent nuclear war led to the cooperation in dialogue between the two super powers.

8.3 Characteristics of the Future World

From the above analysis we can characterise the structure of the future world. An individual in the future world will be much better informed and aware. They will be more powerful, owing to increased knowledge. They will be networked with other individuals and able to form many varieties of organisational structures. However, they will be more pressurised and stressed. Their dissatisfaction levels will be high, their patience low and they will be prone to take extreme measures.

The organisations formed by these individuals will be highly networked. Hierarchies will have to become networks. These organisations will be much more agile. They will have to be collaborative. Though these structures will be prone to disasters because of technological failures – either malfunctioning or human-created – nevertheless it is becoming clear that networking and knowledge will become the core of human existence.

Hence the characteristics of the future world that now looms will be multipolar, rapidly changing, highly networked, knowledge-based and technologically advanced. This world will also, however, be composed of a high percentage of have-nots and disgruntled or dissatisfied individuals or organisations that can network to disrupt the normal functioning of other individuals and organisations.

The solutions to the problems of the security of a nation in such a dynamic and fast-paced world will need to be agile and flexible. Prioritising national security requirements in such a world requires a detailed look at the strategic forces and constraints that will impact the security of a nation in the medium and long term.

8.4 Factors Impacting National Security

In a world that is highly networked in many dimensions, the threats to the security of a nation can come from many directions. These threats may be internal or external. They can be due to economic or geopolitical factors, or consist of civil wars, insurgencies, social unrest or even technological disasters. It is the responsibility of the government to provide appropriate security to its population and protect the national interest.

Let us consider a hypothetical nation existing in the modern world whose characteristics are as explained in the previous sections. This example will explain the methodology and how the strategic decision-making framework described below can help a modern nation to analyse and evaluate multiple factors impacting its security. The decision-makers of the hypothetical nation need to analyse and evaluate their country's priorities with respect to its security from a multitude of multidimensional threats. In the long term this prioritisation will help the national government in taking better decisions. The long-term period, for the analysis shown here, is assumed to be 25 years. The medium-term time period has been assumed to be 10 years.

The strategic decision-making problem is to identify and rank the factors that will affect the national security of a nation in the medium and long terms. Most relevant factors that can directly or indirectly influence the security of a nation need to be identified. The next step after identifying these factors is to juxtapose them in the environment and in the situation the nation may find itself in the medium and long terms. The shifts in the relative priorities of these factors from medium to long term can provide important insights for the strategic planners. The prioritisation of the factors in the long and medium terms can be done using the analytic hierarchy process (AHP) as described in the framework below.

In the medium term the world is still coping up with the demise of bipolarity, although a clearly unipolar world is visible. This is especially so after the military operations in 2003 to remove the Iraqi regime. There may occur a shift towards multipolarity, which would influence economic, geopolitical and technological factors. In the long term, however, the world may stabilise geopolitically into a number of natural poles each of which can defend its interests and stand up to other military powers.

While responding to many threats to its security a country may have to seek help from others through regional military and economic alliances, coalition support and the United Nations (UN). It may have to exploit its geographic situation to aid its defences. Positive steps will have to be taken to contain internal disorders and check insurgency generated by ethnicity and religious fundamentalism, if such a scenario is present.

It can be gathered from this discussion that a nation's security is affected primarily by the following high-level parameters.

1. The world power situation.
2. The eco-commercial situation.
3. Technological factors.
4. Environmental factors.
5. Socio-cultural factors.

The above, along with several sub-factors, can be represented in the form of a hierarchy (see Figure 8.2). The meaning of these sub-factors is explained below.

8.4.1 The World Power Situation (W)

The Shift to a Unipolar Then to a Multipolar World (M)

The future world order is likely to be based on the multipolar model rather than the earlier bipolar world or the present unipolar world. The impact of this shift on the national security of a nation is a major factor that should be considered. The shift to the unipolar world now visible needs to be taken into account in analysing national security factors. If the nation's interests are in conflict with the interests of the superpower of the unipolar world, then how it should shape its security environment in the medium and long terms may become a crucial issue. However, if such interests are not in conflict with the super power then the security implications may be completely different, and if the interests are aligned with the super power then the strategy for national security may be radically different. If the country considers that in the long term multipolarity will be the norm, then the steps taken by the country towards achieving its security objectives may vary.

Regional and Global Military Groupings (G)

Many countries form covert or overt defence alliances that can impact the nation's security in a major way. It is one thing to have a conflict with a single nation and an altogether different matter when the adversary is backed by military alliances. Hence it is of critical importance to look at how these military alliances are likely to evolve in the medium- and long-term scenarios.

The Role of the UN (U)

Despite the not very effective role played by the United Nations in the present unipolar world, the UN continues to be the sole institution where the internationalisation of various issues can be carried out. How the regional issues are played in the UN arena may become crucial from the security point of view of a nation. Hence the role of the UN and the position of the nation in the UN in the medium and long-term time periods needs to be properly understood and responded to on the diplomatic and other fronts.

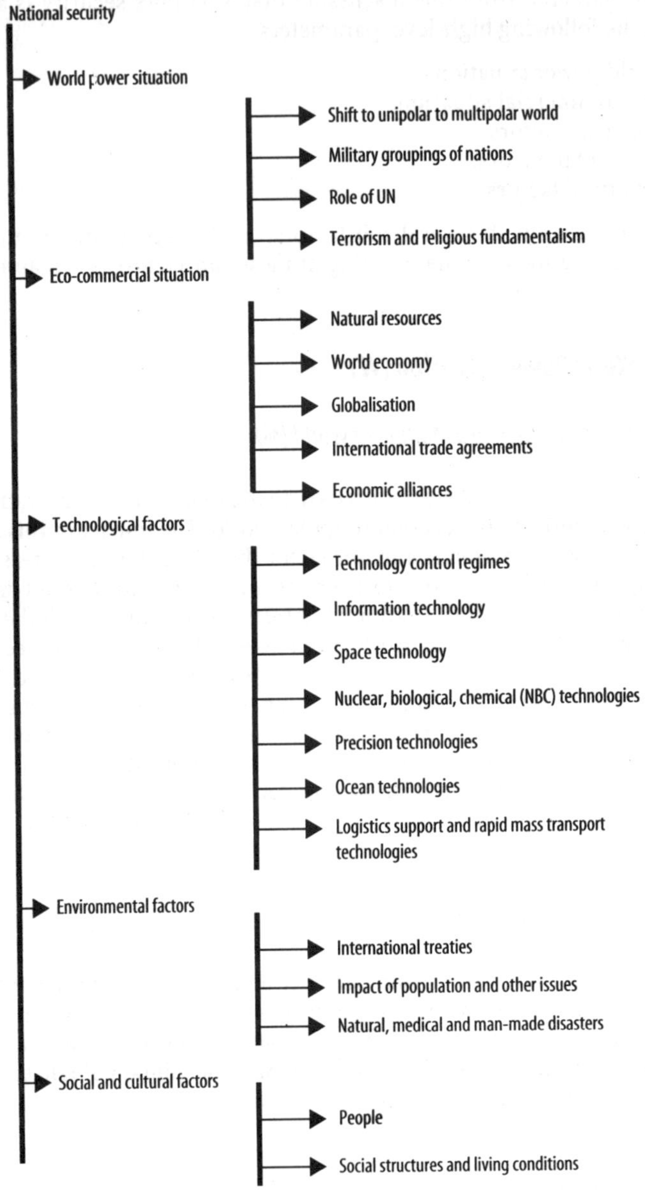

Figure 8.2 Main factors and sub-factors affecting a country's national security.

Terrorism and Religious Fundamentalism (T)

The 11 September 2001 destruction of the World Trade Centre (WTC) showed the growing geopolitical aspect of terrorism. The growing menace has already become a major challenge to world stability and it is of paramount importance to respond to terrorism at all levels and in all dimensions. How the nation fights the terrorist organisations, religious fundamentalism and internal disparities may impact on its own long-term stability.

8.4.2 The Eco-commercial Situation (C)

Natural Resources (N)

From the security point of view, a nation rich in natural resources has to ward off attempts by other countries to take control of these. On the other hand, a country with no such abundance has to look for natural resources elsewhere. The most important natural resource is oil. If a country has sufficient reserves or can meet its growing energy needs then the security effort should be to protect these resources. However, if the oil needs are not met by in-house resources then oil has to be imported. This creates a major potential threat to the eco-commercial situation and in turn the security of the nation.

The World Economy (E)

We are living in a highly connected world. The "butterfly effect", i.e. the fluttering of the wings of a butterfly in China setting off a chain of events that ends with rain falling in New York, is truly and clearly observable in principle in the workings of the world economic system. As the world economy goes through its series of ups and downs, the national economy of a modern nation also goes through major undulations. These have the potential to cause multiple economic crises and even disasters. Any analysis of the security situation of a country should necessarily take the state of the world economy into account.

Globalisation (G)

The opening of the national economy to the outside world economy and various other interactions are creating many new means and markets for private organisations to acquire wealth. The increasing connections between geopolitics, the sociocultural mixing of civilisations, the shared awareness of world events, the growing networks of networked economies, the increasing gap between poor and rich, and the increasing pressures to break trade barriers have always been results of the global village. However, the pace at which these dynamical activities have gone on for the last decade or so really has created a world that is permanently connected. The new technological infrastructure led by internet communication between people around the world is having a hitherto unimaginable impact on general awareness, economic conditions and requirements, social conditions and political awareness, all with serious repercussions on the security of a nation.

International Trade Agreements (T)

Although globalisation is happening at a rapid pace, there have been various attempts to control and moderate its speed, which sometimes manifest themselves as international trade agreements. Such agreements are important in breaking trade barriers and also in protecting the national interests of the member countries. As an example, the TRIPS treaty is an attempt to harmonise the intellectual property (IP) laws of constituent countries so that trade between them can carry on without any fear of IP infringement.

Economic Alliances (A)

Inclusion or exclusion from major economic groupings may occur at both regional and global levels. These economic alliances can create competitive disadvantages for the nation that does not want to be a member. For a nation, it may be beneficial to be part of these alliances. A thorough analysis is needed to look at the costs and benefits from a long-term perspective. However, the security implications of these alliances are obvious.

8.4.3 Technological Factors (T)

Technology Regimes (T)

Crucial technologies, relevant to the development of a society, have a dual capability in that they can be used to make better weapons and weapons systems. There is a tendency to control the proliferation of technologies around the world. This controlling itch among the technologically advanced nations has resulted in many technology regimes filtering out nations that are lagging in the technology race. A nation's security needs will be impacted by which side of such regimes the nation lies. The technologically have-not nations will have to take different measures to catch up with these regimes.

Information Technology (IT)

Information technology has become such a critical factor in everyday life that the very existence of a nation can be influenced by it. The security, economy and social life of a modern nation is heavily dependent on IT.

Aerospace Technologies (A)

The final frontier has become a leading area for the creation of technologies that give continuous surveillance, control (e.g. geographical positioning systems (GPS)) and precision-destructive capabilities. The aerospace technologies are the key component of technology factors that will influence the security of a nation.

Nuclear Biological And Chemical Technologies (NBC)

The effect of progress in NBC technologies on the security of a nation will definitely be considerable.

Precision Industrial Technologies (PIT)

The precision technologies at the nanoscale and at the heavy industrial scale will impact how future products are developed. Advances in these technologies will also be critical in the design of new weapon systems.

Ocean Technologies (O)

Ocean technologies and the ability to control oceanic areas in the world through advancements in technology will be critical to a nation's security.

Rapid Mass Transport Technologies (M)

The effect of progress in rapid mass transport systems will be considerable for providing a nation with the strategic ability to move men and material rapidly across large geographical distances during crises and disasters.

8.4.4 Environmental Factors (E)

International Treaties (T)

International treaties on ecology and the management of the environment can impact a nation's security and industrial growth.

Impact of Population and Other Issues (M)

The effect of human activities on ecology and the environment include the ability to handle pollution and population explosion.

Natural Disasters (D)

The ability to handle natural disasters will be a critical factor in taking care of future uncertainties. By far the most important element is the ability to take care of natural calamities including medical disasters. This ability assumes greater importance in the wake of the increasing threat from the non-state actors (e.g. terrorist groups) to disrupt the normal functioning of a nation by the use of dirty nuclear, biological and chemical (NBC) weapons.

8.4.5 Socio-cultural Factors (S)

People (P)

Population growth and migration affect social relations, cultural dynamics, economic conditions, industrial development, and generic potential of population.

Social Structures and Living Conditions (D)

Literacy, income distribution, social security, etc. affect social relations, e.g. in disputes due to land tenure, the treatment of the sick and the poor, exploitation, caste, etc.

After identifying these factors and organising them into the hierarchical levels as given in the text and shown in Figure 8.2, the next step is to elicit pairwise comparisons on these parameters from various experts. These experts have to be given clear, unambiguous guidelines on how to fill up the relevant forms to obtain the proper comparisons. The experts should be drawn from various fields of security decision-making. They can be from academia, the military, the security sector, economic institutions, industry, and government organisations. The inputs obtained separately from various experts need to be analysed for inconsistencies, using the procedure described in Chapter 2. With the highly inconsistent (with consistency ratio >0.20) matrices eliminated or reworked in consultation with the experts, the comparison matrices need to be combined into single matrices using the procedures described in that chapter. Let us assume that the final matrices combining all these comparisons come out to be as shown in Section 8.5 below. These matrices are used to elicit the relative importance of factors affecting national security in the medium and long terms and can be employed to define national security strategy for the long-term and -scenarios.

8.5 Priorities of Factors in the Medium and Long Terms

The above procedure leads to 12 matrices, 6 belonging to the medium and 6 to the long term. Tables 8.1 and 8.2 show the comparisons of 5 main factors, i.e. the world

Table 8.1 Main factors (medium-term).

	P	*C*	*T*	*N*	*S*
P	1.000	1.401	1.528	6.544	4.663
C	0.713	1.000	1.838	5.431	4.210
T	0.654	0.544	1.000	5.078	5.165
N	0.153	0.184	0.197	1.000	0.903
S	0.214	0.238	0.194	1.107	1.000
Importance	0.347	0.301	0.239	0.052	0.061

Consistency ratio = 0.015.

Table 8.2 Main factors (long-term).

	P	*C*	*T*	*N*	*S*
P	1.000	0.245	1.100	2.188	1.372
C	4.078	1.000	2.704	7.071	4.884
T	0.909	0.370	1.000	5.265	4.126
N	0.457	0.141	0.190	1.000	0.818
S	0.728	0.205	0.242	1.222	1.000
Importance	0.146	0.486	0.229	0.059	0.080

Consistency ratio = 0.036.

power situation (P), the eco-commercial situation (C), technological factors (T), environmental factors (N) and social and cultural factors (S). The last row in these matrices shows the relative importance of the 5 main factors obtained by normalising the right eigenvector corresponding to the principal eigenvalue of the matrices. The consistency ratio of these matrices is also shown below the tables.

The Kiviat chart corresponding to the medium-term impact of these factors on national security is shown in Figure 8.3. As is evident from these figures, experts believe that the highest level, the world power situation, and how it stabilises will have the most important influence (relative importance = 34.7%) on the security of the nation in the medium term. However, the eco-commercial situation is not far behind (relative importance = 30.1%) and in fact quite important when looking at the influence on national security. Also in the medium term, a major influence will be how the nation taps the technological factors and advances taking place in a variety of technologies for enhancing its security (relative importance = 23.9%).

The Kiviat chart corresponding to the long-term impact of the main factors on national security is shown in Figure 8.4. It can be seen that in the long-term scenario the overriding factor becomes the eco-commercial situation. The experts believe the eco-commercial situation will be the major influence on national security, with a relative importance of 48.6%. The impact of the world power situation in the long

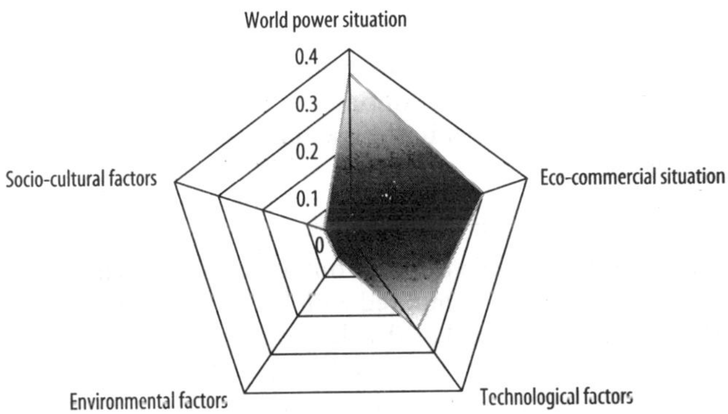

Figure 8.3 First-level factors affecting national security in the medium term.

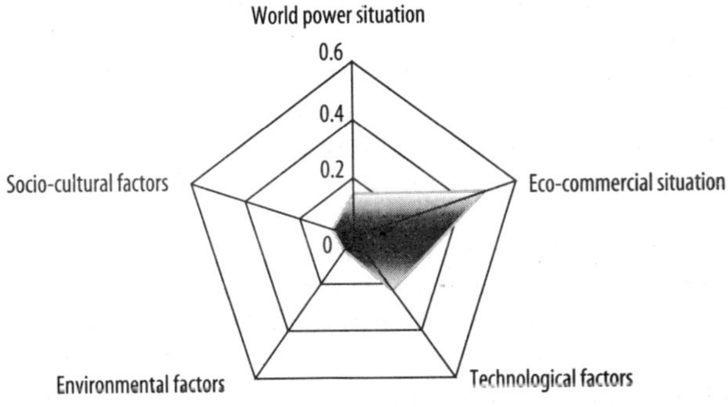

Figure 8.4 First-level factors affecting national security in the long term.

Table 8.3 Sub-factors under the world power situation (medium-term).

	M	*G*	*U*	*T*
M	1.000	1.100	3.100	0.800
G	0.909	1.000	2.350	0.650
U	0.323	0.426	1.000	0.135
T	1.250	0.415	7.407	1.000
Importance	0.289	0.245	0.086	0.379

Consistency ratio = 0.037.

Table 8.4 Sub-factors under the world power situation (long-term).

	M	*G*	*U*	*T*
M	1.000	0.900	1.550	0.900
G	1.111	1.000	1.750	1.010
U	0.645	0.571	1.000	0.553
T	1.111	0.415	1.808	1.000
Importance	0.270	0.301	0.171	0.258

term will decrease compared with the technological factors. The relative importance of technological factors will continue to be almost same (i.e. 22.9%) in the long term, compared with 23.9% in the medium term. It is believed that over the long term, the world power situation will stabilise and its influence on national security will reduce (relative importance = 11.6%). The relative importance of social and cultural factors will improve to 8.0% in the long term compared with 6.1% in the medium term, while the environmental factors will increase in relative importance to a value of 5.9% compared with 5.2% in the medium term.

Each of the five factors is further divided into various sub-factors. Following a procedure similar to that observed for the main factors, analysis is carried out of the impact of the sub-factors on the main factors. However, what the experts want to know at this stage is the relative importance of one sub-factor over another with respect to its contribution to the main factor in the medium-term and long-term scenarios. Tables 8.3 and 8.4 show the matrices corresponding to the sub-factors contributing to the world power situation in the medium term and long term respectively. The four sub-factors contributing to the world power situation as described earlier are a shift from a unipolar to a multipolar world (M), regional and global military groupings (G), the role of the United Nations (U) and terrorism and fundamentalism (T). The last rows in these tables show the relative importance of sub-factors with respect to their contribution to the main factors in the medium and long terms.

The Kiviat chart corresponding to the medium-term impact of sub-factors contributing to the world power situation is shown in Figure 8.5. As is evident there, experts believe that the world power situation will be impacted most by terrorism and fundamentalism (relative importance = 37.9%) in the medium term. The shifting power dynamics from a unipolar to a multipolar world will also have a major influence on the world power situation (relative importance = 28.9%). The

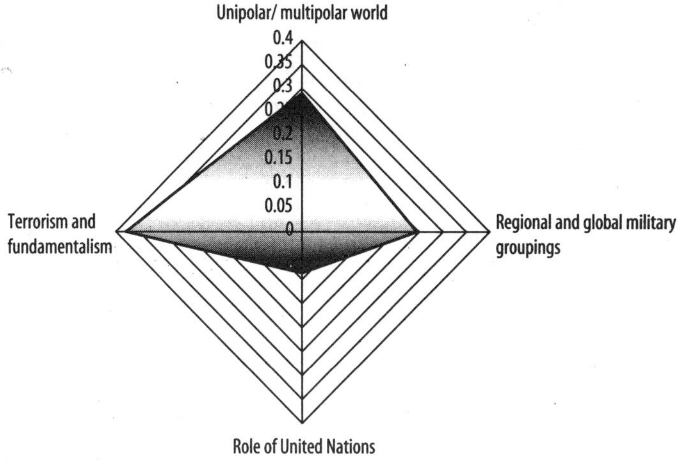

Figure 8.5 Sub-factors corresponding to the world power situation in the medium and long terms.

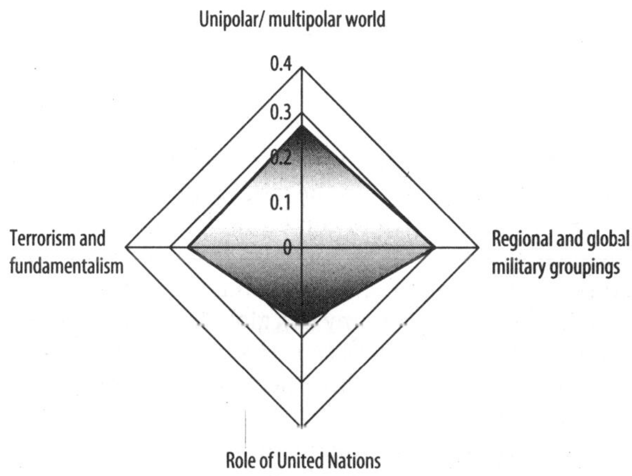

Figure 8.6 Sub-factors corresponding to the world power situation in the long term.

regional and global military groupings also seem to have a very important role to play (relative weight = 24.5%). The role of the UN in the medium term is believed by the experts to have very low relative importance in contributing to world power situation (relative weight of 8.6%).

According to the experts however, in the long term the situation will change. The Kiviat chart in Figure 8.6 shows that the experts believe that global and regional military alliances will make a major contribution to the world power situation (relative importance = 30.1%). This is a clear indication that the world will move towards multipolarity (relative importance = 27.0%). This shift will come about through military alliances, rather than the rise of any individual competitor. Terrorism and fundamentalism will continue to be a major challenge in the long term as well (relative importance = 25.8%). However, the role of the UN, as a

Table 8.5 Sub-factors under the eco-commercial situation (short-term).

	N	*E*	*G*	*T*	*A*
N	1.000	4.000	0.191	0.299	0.275
E	0.250	1.000	0.330	0.330	0.330
G	5.233	3.000	1.000	1.000	1.000
T	3.344	3.000	1.000	1.000	0.670
A	3.637	3.000	1.000	1.500	1.000
Importance	0.189	0.114	0.186	0.236	0.275

Consistency ratio = 0.006.

Table 8.6 Sub-factors under the eco-commercial situation (long-term).

	N	*E*	*G*	*T*	*R*
N	1.000	4.000	0.330	0.508	0.508
E	0.250	1.000	0.330	0.330	0.330
G	3.000	3.000	1.000	1.000	1.000
T	1.968	3.000	1.000	1.000	1.000
R	1.968	3.000	1.000	1.000	1.000
Eigenvector	0.204	0.042	0.178	0.186	0.389

Consistency ratio = 0.025.

contributor to world power situation, will become more important (relative weight = 17.1% compared with 8.6% in the medium term).

Tables 8.5 and 8.6 show the matrices corresponding to sub-factors contributing to the eco-commercial situation in the medium term and long term respectively. The five sub-factors contributing to eco-commercial situation are the availability of natural resources (N), the world economy (E), globalisation (G), international trade agreements (T) and economic alliances (A). The last rows in these tables show the relative importance of sub-factors with respect to their contribution to the main factors in the medium term and long term.

The Kiviat chart corresponding to the medium-term impact of sub-factors contributing to the eco-commercial situation is shown in Figure 8.7. As is evident from these figures, the experts believe that the eco-commercial situation will be impacted most of all by economic alliances (relative importance = 27.5%) and international trade agreements (relative importance = 23.7%). The availability of natural resources (relative importance = 18.9%) and globalisation (relative importance = 18.3%) will have an almost similar impact on the eco-commercial situation. The world economy in the short term will impact the eco-commercial situation in the short term to the least extent (relative importance = 11.4%).

The Kiviat chart corresponding to the long-term impact of sub-factors contributing to the eco-commercial situation is shown in Figure 8.8. The experts believe that the eco-commercial situation will be impacted most by the economic alliances (relative importance = 38.9%). However, the relative importance of natural resources has increased considerably in the long term, as these resources become more and more scarce (relative importance = 20.4%). International trade agreements (relative importance = 18.6%) have decreased in relative importance

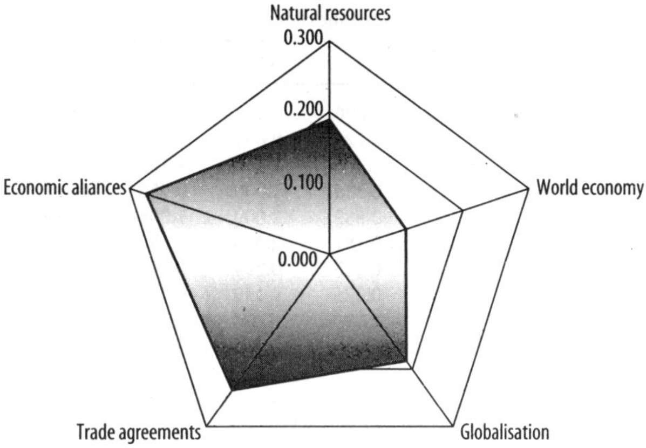

Figure 8.7 Sub-factors corresponding to the eco-commercial situation in the medium term.

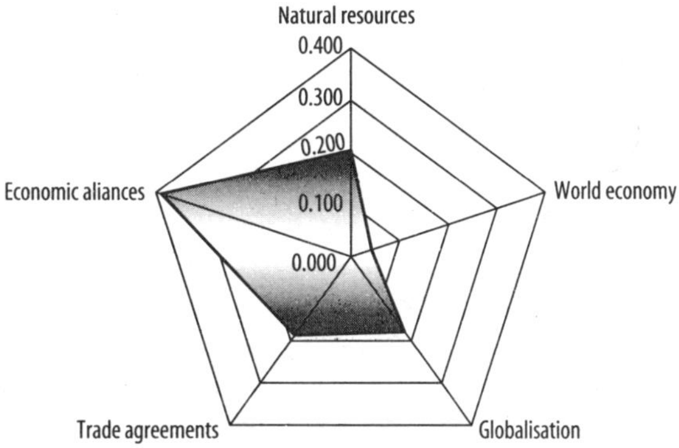

Figure 8.8 Sub-factors corresponding to the eco-commercial situation in the long term.

from the point of view of their impact on the eco-commercial situation. The impact of globalisation (relative importance = 17.8%) has remained more or less the same. The world economy in the long term will impact the eco-commercial situation to the least extent, with a relative importance of 4.2%.

Tables 8.7 and 8.8 show the matrices corresponding to sub-factors contributing to the technological factors in the medium term and long term respectively. The seven sub-factors contributing to technological factors impacting national security are technology control regimes (T), information technology (IT), aerospace technologies (A), nuclear, biological and chemical technologies (NBC), precision industrial technologies (PIT), ocean technologies (O) and rapid mass transport technologies (M). The last rows in these tables show the relative importance of sub-factors with respect to their contribution to the main factors in the medium term and long terms.

The Kiviat chart corresponding to the medium-term impact of sub-factors contributing to technology factors is shown in Figure 8.9. Experts believe that sub-factors contributing to technological factors impacting national security will be the adoption of IT (relative importance = 24.4%) and aerospace technologies (relative

Table 8.7 Sub-factors under technology (short-term).

	T	*IT*	*A*	*NBC*	*PIT*	*O*	*M*
T	1.000	5.000	5.000	5.165	5.525	5.000	5.000
IT	0.200	1.000	0.229	0.654	0.394	2.370	5.165
A	0.200	4.360	1.000	2.954	1.125	5.525	6.540
NBC	0.194	1.528	0.338	1.000	0.356	0.422	4.360
PIT	0.181	2.536	0.889	2.809	1.000	0.725	5.525
O	0.200	0.422	0.181	2.370	1.380	1.000	3.270
M	0.200	0.194	0.153	0.229	0.181	0.306	1.000
Importance	0.146	0.244	0.221	0.139	0.110	0.085	0.055

Consistency ratio = 0.024.

Table 8.8 Sub-factors under technology (long-term).

	T	*IT*	*A*	*NBC*	*PIT*	*O*	*M*
T	1.000	0.200	0.175	0.207	0.204	0.525	1.380
IT	5.000	1.000	0.390	1.246	0.338	0.803	1.904
A	5.720	2.540	1.000	4.830	1.380	2.140	3.270
NBC	4.829	0.803	0.207	1.000	0.356	0.582	1.552
PIT	4.904	2.954	0.725	2.809	1.000	1.125	1.933
O	1.904	1.246	0.467	1.719	0.889	1.000	0.422
M	0.725	0.525	0.306	0.644	0.517	2.370	1.000
Importance	0.047	0.352	0.260	0.090	0.072	0.085	0.094

Consistency ratio = 0.041.

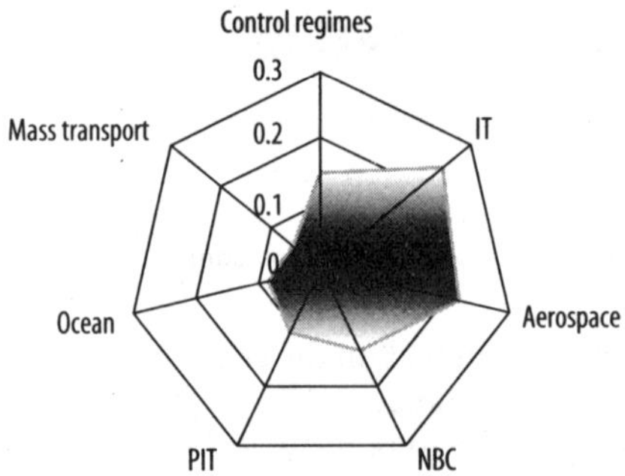

Figure 8.9 Importance of technological sub-factors in the medium term.

importance = 22.1%). Technology control regimes (relative importance = 14.6%) and NBC technologies (relative importance = 13.9%) will also play a major role in the medium term. The relative contribution of precision industrial technologies in the medium term will be 11.0%, while ocean technologies and rapid mass transport technologies will contribute 8.5% and 5.5% respectively.

The Kiviat chart corresponding to the long-term impact of sub-factors contributing to technology factors is shown in Figure 8.10 below. Information technology (IT) continues to be a major contributor in the long run (relative importance = 35.2%), followed by aerospace technologies (relative importance = 26.0%). It can be seen that the relative contribution of NBC technologies has actually gone down to 9.0%,

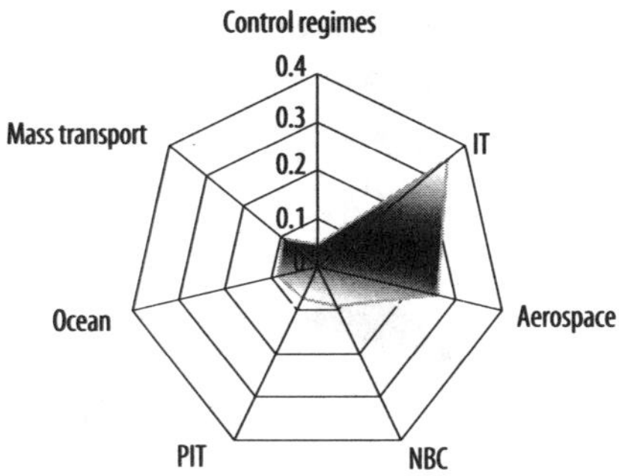

Figure 8.10 Importance of technological sub-factors in the long term.

Table 8.9 Sub-factors under environment factors (medium-term).

	T	*P*	*D*
T	1.000	0.902	0.180
P	1.109	1.000	0.270
D	5.556	3.704	1.000
Importance	0.138	0.169	0.693

Consistency ratio = 0.009.

Table 8.10 Sub-factors under environment factors (long-term).

	T	*P*	*D*
T	1.000	0.750	0.122
P	1.333	1.000	0.330
D	8.197	3.030	1.000
Importance	0.109	0.184	0.707

Consistency ratio = 0.048.

while the relative contribution of rapid mass transport technologies has gone up to 9.4%. It is evident that the experts believe that the rapid movement of people and material will in the long run be a critical capability that the nation should develop.

Tables 8.9 and 8.10 show matrices corresponding to sub-factors contributing to the environmental factors in the medium term and long term respectively. The three sub-factors contributing to environmental factors impacting national security are international treaties (T), the impact of population growth (P) and disasters (D). The last rows in these tables show the relative importance of sub-factors with respect to their contribution to the main factors in the medium term and long term.

The Kiviat charts corresponding to the medium-term and long-term impact of sub-factors contributing to environmental factors are shown in Figures 8.11 and 8.12, respectively. Experts believe that disaster-handling capability will be the key in the medium and long term for the nation (relative importance of almost 70% in the medium and long terms).

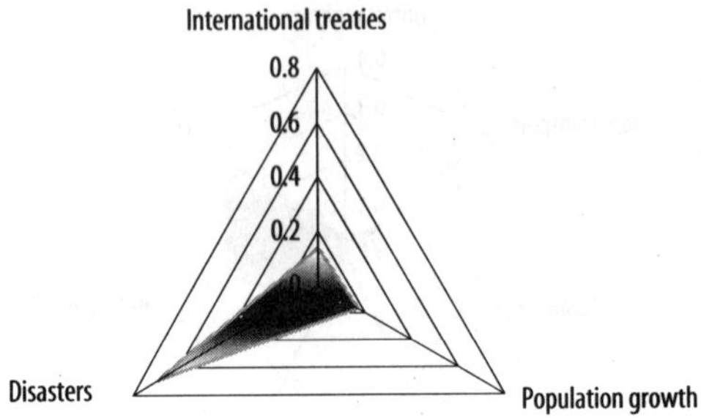

Figure 8.11 Importance of environmental sub-factors in the medium term.

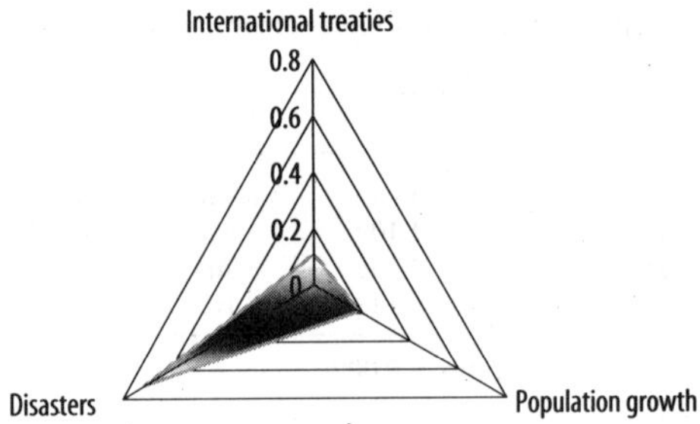

Figure 8.12 Importance of environmental sub-factors in the long term.

Table 8.11 Sub-factors under socio-cultural factors (short-term).

	P	*L*
P	1.000	3.225
L	0.310	1.000
Importance	0.763	0.237

Table 8.12 Sub-factors under socio-cultural factors (long-term).

	P	*L*
P	1.000	1.202
L	0.832	1.000
Importance	0.546	0.454

Tables 8.11 and 8.12 show the matrices corresponding to sub-factors contributing to socio-cultural factors in the medium term and long term respectively. The two sub-factors contributing to socio-cultural factors impacting national security are people (P) and social structures and living conditions (L).

Table 8.13 Overall importance of factors affecting national security in the medium and long terms.

Medium term	%	*Long term*	%
Terrorism	13.15	Economic alliances	18.91
Shift to unipolar to multipolarity	10.03	Natural resources	9.91
Military groupings	8.50	International trade agreements	9.04
Economic alliances	8.28	Globalisation	8.65
International trade agreements	7.10	IT	8.06
IT	5.83	Aerospace technologies	5.95
Natural resources	5.69	Military groupings	4.39
Globalisation	5.60	People	4.37
Aerospace technologies	5.28	Disasters	4.17
People	4.65	Shift to unipolar to multipolarity	3.94
Disasters	3.60	Terrorism	3.77
Technology regimes	3.49	Living conditions	3.63
World economy	3.43	Role of UN	2.50
NBC technologies	3.32	Mass transport technologies	2.15
Role of UN	2.98	NBC technologies	2.06
Precision industrial technologies	2.63	World economy	2.04
Ocean technologies	2.03	Ocean technologies	1.95
Living conditions	1.45	Precision industrial technologies	1.65
Mass transport technologies	1.31	Population growth	1.09
Population growth	0.88	Technology regimes	1.08
Environment treaties	0.72	Environment treaties	0.64
Total*	**99.95**	**Total***	**99.95**

*But for truncation errors the sum would be 100.

In the overall analysis, the relative importance of various sub-factors has been evaluated by multiplying the relative importance the sub-factor with the relative importance of the relevant main factor. The total picture that emerges from this analysis is shown in Table 8.13.

In the medium term (see the Kiviat chart in Figure 8.13) the security of the nation will be impacted by terrorism (relative importance = 13.15%) and the power shift to a unipolar and perhaps to a multipolar world (relative importance = 10.03%). Military groupings at the global and regional level (relative importance = 8.50%), economic alliances (relative importance = 8.28%) and international trade agreements (relative importance = 7.10%) will also influence national security to a considerable extent. In the medium term, the nation should definitely look at information technology (IT) as a possible solution to some of the security requirements (relative importance = 5.83%), try to enhance and tap natural resources (relative importance = 5.69%), take care of globalisation and its impact on its security (relative importance = 5.60%) and develop capabilities in aerospace technologies (relative importance = 5.28%). Developing the quality of people, taking care of systems and operations for disaster avoidance and management, technology regimes and world economy are the next issues of importance in the medium term.

In the long term (see the Kiviat chart in Figure 8.14), however, experts believe it is in economic growth that the security of the nation lies. The most important strategic input is to develop and take care of economic alliances (relative importance = almost 19%). Tapping natural resources (relative importance = 9.91%), international trade agreements (relative importance = 9.04%) and globalisation (relative importance = 8.65%) are also major guidelines in the long run. It can be seen that in the long run also, IT and aerospace technologies each continue to be a major contributor to national security. Developing the quality of people and

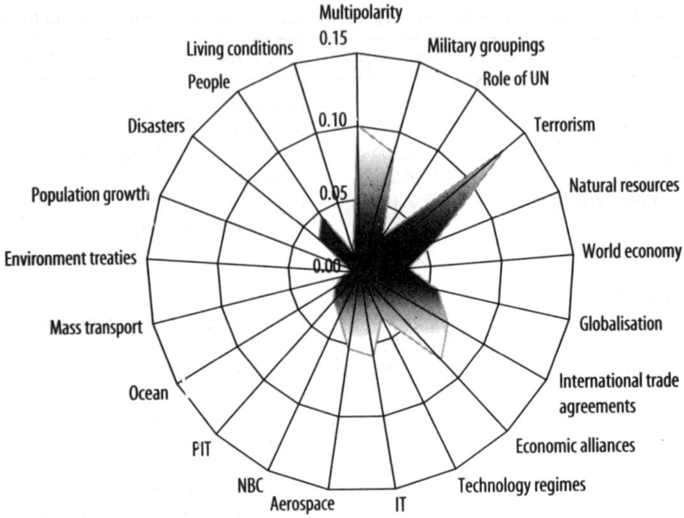

Figure 8.13 Factors affecting national security in the medium term.

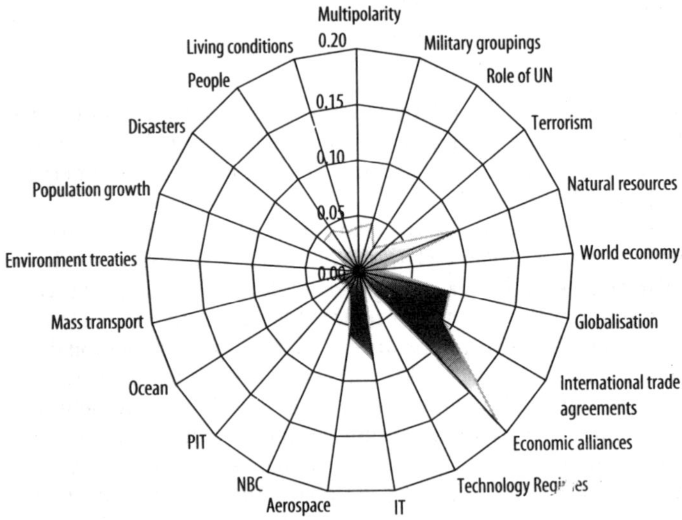

Figure 8.14 Factors affecting national security in the long term.

creating disaster management systems are other prime factors contributing to national security in the long run.

8.6 National Security Strategic Guidelines for the Nation

From the above analysis, it is clear that in the medium term the biggest impact on national security will be through terrorism, power shifts in the world, military groupings and economic alliances. In the long term, however, economic factors take precedence. Hence the following generic strategic guidelines can be carved out for the nation after this analysis.

- The nation should develop economic alliances with other countries or groups of countries. These alliances may be bilateral or multilateral. It should keep an eye on covert or overt military alliances developing all over the world. If necessary, it should gather intelligence on how these relationships are developing, including any military dimension or links to non-state actors such as terrorist groups. It should keep track of how multiple power poles are evolving around the world. Also, it should continuously evaluate the memberships of these groups and whether it is in the national interest to join hands in these economic, military or technology alliances.
- Globalisation is the key to the future. However, the pace at which globalisation of the nation proceeds should be controlled in a methodical manner. Embracing information technologies (IT) is the key requirement for the economic uplift of the nation. Hence IT should be developed and embraced in all walks of life and also at all levels of government, industry and security organisations.
- Investment in aerospace technologies seems to be the major requirement for taking care of the security of the nation in the long run. Hence efforts should be made to develop these technologies, taking care of technology regimes that invariably will try to control these technologies.
- Disaster and crisis management (see Chapter 9) infrastructure, technologies and organisations will need to be enhanced and properly shaped up for quick responses to the emerging crises and disasters at various times and in different scenarios.
- People will need to be developed. Educational and employment opportunities will need to be enhanced. Living and social conditions will require to be raised.

These are very general guidelines. However, the detailed analysis shown in this chapter can be used in real life to understand the strategic environment the country will be in, in the medium and long terms.

References

1. Toffler A and Toffler H (1998) War and Anti-war Survival at the Dawn of 21st Century. Warner Books.
2. Toffler A (1990) Powershift – Knowledge, Wealth and Violence at the edge of 21st Century.
3. Anton PS, Silberglitt R and Schneider J (2001) The Global Technology Revolution: Bio/Nano/Material Trends and their Synergies with Information Technology by 2015, MR-1307-NIC. Rand Corporation, Santa Monica, CA.
4. Paschall R (1990) LIC 2010 – Special Operations and Unconventional Warfare in the Next Century. Brassey.
5. Ronfeldt D and Arquilla J (2001) Networks, netwars and the fight for the future. First Monday, 6, Number 10.

9 Managing Crisis and Disaster

9.1 Crises and Disasters

In June 2001, three months before 11 September, a tabletop exercise was held at Andrews Air Force Base, USA. It simulated the intentional release of smallpox in three US cities. In the two weeks of the simulated game, 1000 deaths were projected, with 15 000 people infected. The exercise was named "Dark Winter" [5]. It showed what the consequences of a crisis can be if it is not detected and responded to within the time available. Three months later disaster struck New York, reducing the twin towers of the World Trade Center to dust and rubble.

In today's world, crisis and disaster management is a major requirement for all business, defence and government organisations [1]. How efficiently any crisis or disaster is handled depends upon the decision-making capabilities and resources available. Since it is understood that no organisation can live without ever facing a crisis, there is a requirement to manage crisis situations efficiently. Further, even when a disaster strikes, managing the consequences – rebuilding the pieces – is a major challenge for the organisation. Waiting for disasters and crises and handling each as the situation unfolds can lead to grave consequences for the organisation. Hence it is imperative that strategic plans and operational plans be developed, resources deployed and organised with proper infrastructure and decision-makers trained to take decisions during crisis situations. At the organisational level there is a need to have a strategic decision-making framework to cater for likely crisis and disaster situations.

Most businesses all over the world are bound to face crisis situations. In some small and medium businesses handling a business crisis is almost a regular activity. The crisis can be anything from a natural disaster to a sexual harassment legal case. It may come in the form of product defects and recalls, strikes and labour unrest, industrial accidents and hostile takeover attempts. Crises usually result in adverse consequences and if not managed can be disastrous for the business organisation. At the government level, crisis and disaster situations can occur not only because of natural disasters which include earthquakes, floods, hurricanes, wild fires, etc. but also owing to various emergencies such as armed conflicts, insurgency, terrorist activities and civil unrest. Managing such disasters and crises is the prime responsibility of all organisational units.

A disaster is defined as an event – natural or man-made – which impacts with such severity that the affected community, organisation or group has to respond by taking exceptional measures. Although in some places a disaster usually means a natural event (e.g. an earthquake or floods), in this chapter we will consider disaster at the generic level. If the disaster strikes without warning, a sudden earthquake for example, the organisation has to rebuild the destroyed elements. However, many disasters provide enough warning of their coming. This is the period of crisis. In this period there are options for the organisation involved to decide upon the course of action which can pull back the situation from the imminent threat. There is thus a clear case for any organisation to put systems in place which will prevent a situation reaching even a crisis point. This is the phase of crisis avoidance. It and the specific responses during it are defined by the type of disaster and the situation the organisation is in.

9.2 Classification of Disasters

A disaster can happen in any field. It implies a major setback to the organisation and sometimes the complete annihilation of structures and people. The specific nature of a disaster may vary from one area to another. Figure 9.1 shows the types of disaster and their long-term and short-term impacts. The warning time given for each of these is also plotted. Three broad categories of disasters have been considered here – natural disasters, business disasters and weapons of mass destruction (WMD) disasters. Evaluating these generic classes of disasters in terms of long-term impact, short-term impact and warning time, one can see that the business and natural disasters can have up to high immediate impact. The long-term impacts of business disasters are medium to high. However, the WMD disasters can impact an organisation to the maximum in both the immediate period and the long-term. Looking at the warning time dimension, it can be seen that business and natural disasters can give a warning time up to medium, whereas the WMD disaster's warning time is very low. With extremely high immediate and long-term impacts and an extremely low warning time, WMD disasters are thus very difficult to manage.

Natural disasters include earthquakes, floods, cyclones, landslides, drought, famine, avalanches, volcanic eruptions, dust storms, tornadoes and hurricanes,

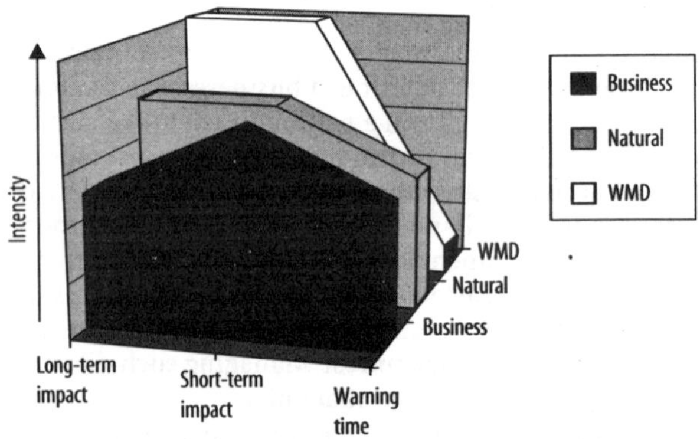

Figure 9.1 Type of disasters and their impact.

wild fires, etc. Though the medical disasters which include the rapid spread of communicable diseases are not strictly speaking natural, we will consider them here in the natural disaster category. The spread of acquired immune deficiency syndrome (AIDS) is a case in point. The industrial disasters such as the accidental release of poisonous gases from an industrial plant are considered in the business disaster category, though they will impact the community where the industrial plants are located. Air crashes, shipwrecks and oil spills are also included among business-related disasters, although the government agencies have to manage these disasters because of their impact on the community. In the 1991 Gulf War, Iraq threatened to burn nearly 800 oil wells in Kuwait. The impact of this crisis was compared to that of a nuclear conflict, and there were talks of the whole Middle East region facing a situation similar to a "nuclear winter". Fortunately, whatever oil wells were set on fire could not continue for long. Such crises and disasters fall between the categories "natural" and "business".

By far the most critical disasters are those in the category of weapons of mass destruction (WMD). WMDs are the greatest threat to peace and stability internationally and even to human existence. The three categories of WMDs, i.e. nuclear, biological and chemical (NBC) weapons, although deplored by almost the whole world, still find enough strategic or deterrence value to be possessed by many states. The state actors with WMD capability are increasing. The increasing proliferation of state actors increases the possibility of these weapons falling into the hands of non-state actors, i.e. terrorists and guerrilla groups pursuing the control of territory and/or ideological or religious ends. With the increase in the possibility of the possession and usage of WMD by non-state actors, it is of the utmost importance that proper organisations, response centres, resources and personnel should be set up, trained and ready to respond to such threats.

The increasing pace of industrialisation also indicates that in future NBC incidents with the possibility of leading to disasters may occur from the proliferation of nuclear power plants, large chemical plants or even biotechnology laboratories. Although these plants adopt many safety measures, no system can however be entirely fail-safe; if some non-state actor wants to create a WMD incident, it can go to such a plant and execute its plans.

To respond to a disaster, an organisation needs to understand how disasters actually happen. The stages through which a situation can reach a disaster point should be an important input to any strategic plan of counteraction.

9.3 Disaster Stages

Figure. 9.2 shows the three stages or phases in any disaster management scenario, namely crisis avoidance, crisis management and disaster management. As can be seen from the graph, the stakes increase exponentially as the situation progresses through the three phases, culminating in the disaster phase where the stakes are very high.

It can be observed from the figure that the crisis avoidance phase is quite long, indicating that usually the situation continues to be peaceful in most scenarios. However, a crisis when it occurs has to be responded to in a very short span of time, as shown in the narrower box of the crisis management phase. Therefore, it makes

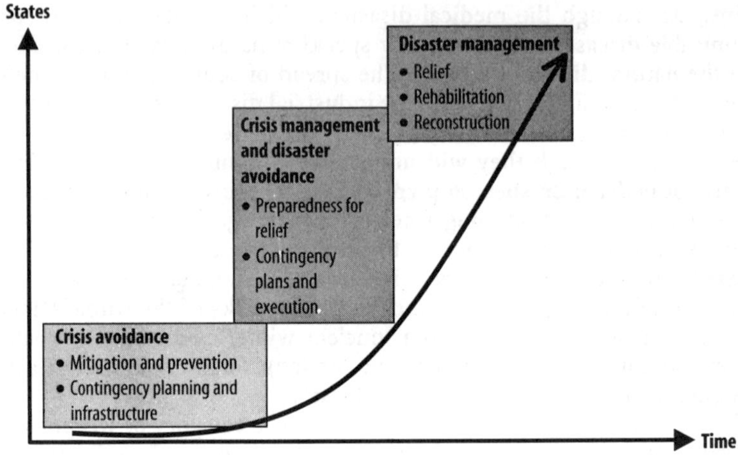

Figure 9.2 Stages in disaster management.

sense to invest resources and energy in avoiding crisis situations in the first phase, then responding during the management phase. The crisis management phase assumes much more importance, as the demands for decision-making during this phase are extreme. The decision-maker is under tremendous pressure to balance all the constraints and it is likely that if their responses are not automatic they will fail. Hence training of such decision-makers based on certain simulated conditions during the earlier phase is essential. If disaster does strike there will be tremendous loss all around; the organisation, community or nation will have to rebuild everything. This process of managing a disaster is a long one. It takes an enormous amount of time and resources to bring relief to the community, rehabilitate people and reconstruct the organisation. The stakes are really high, as the cost of not managing a disaster is the certain destruction of the organisation, the community or even the nation.

During the crisis avoidance stage there is no indication of any impending disaster, but there is a general idea about the possibility of a disaster situation. This may be from past data, from experience or from new developments. Common sense tells us that preventing a crisis is far safer than trying to escape from it. A risk condition is defined as any action or condition which impacts a situation in such a way that it may lead to a crisis and/or disaster. The risk condition or conditions should be identified, controlled and responded to during this stage. Also during this stage the organisations and decision-makers concerned should see that possible strategies for disaster prevention and mitigation are generated, contingency plans for the response developed and resources earmarked.

When crisis avoidance fails, the situation develops into a crisis, i.e. there is a clear warning of a likely impending disaster; an example would be a hostage situation, when somebody has captured a group of people and is threatening to kill them if certain demands are not met. This is clearly a crisis situation. A crisis is defined as a situation which has a very high probability of leading to a disaster in a very short time period. During a crisis, the crisis management/disaster avoidance phase starts. In order to avoid disaster, alternative contingency plans need to be evaluated or developed, chosen and executed in a compressed time-period. This stage requires quick decision-making and lightening responses.

If crisis management fails then the situation develops into a disaster. Once disaster has struck there comes a need to bring relief, rehabilitation and reconstruction to the affected community, group or business. This is the phase of disaster management. This phase places considerable demands on the energy and resources of all the stakeholders. A similar framework has been described in the US Federal Response Plan [7]. This has a crisis management phase and also what is called the consequence management phase. The latter is similar to the disaster management phase described in the current framework.

9.4 A Systematic Approach to Disaster Management

A strategic approach to responding to crisis and disaster situations is proposed here. The approach calls for the identification of various factors and/or parameters that affect the response capability of the organisation in each of the three phases already outlined, i.e. crisis avoidance, crisis management and disaster management.

9.4.1 Factors Relevant to Crisis Avoidance

Major requirements related to crisis avoidance include surveillance and intelligence gathering, forecasting, dedicated organisation, training, developing technological resources, documentation and databases, crisis management plan generation, public awareness and countering the conditions that may lead to disasters or crisis.

Surveillance and Intelligence Gathering

This activity requires continuous monitoring of the situation. It needs a definition of those situation indicators and threshold values that will indicate that the situation may develop into a crisis or later disaster. Such thresholds can be used to raise alarms. A plan to organise and generate such a system of monitoring should be created and executed. Also, it is important to identify situations that need to be analysed comprehensively.

Forecasting

Identifying all plausible and non-plausible future situations is an almost impossible task. However, there is a requirement to cover comprehensively as many such situations as possible. This requires looking into the future, something that may be extremely difficult. However, what-if analysis, scenario writing and analysis, historical data analysis, and in some situations cartographic analysis are some of the possible techniques that can be employed.

Organization with Unambiguous Accountability

A central organisation is needed to coordinate the activity of crisis avoidance as a whole. This organisation should be well connected to other organisations such as the intelligence agencies, the police department, the office of foreign affairs, the media etc. Also, each department and organisation should have an established, predefined responsibility and accountability to take care of various activities during every stage of crisis avoidance, crisis management and disaster management.

Training

Three categories of people are extremely important during crisis and disaster management. These are the decision-makers, the first responders and the professional executors. Decision-makers include the members of the crisis management organisation. First responders are agencies and organisations that are going to bear the initial brunt of the crisis situation. These may include local police and other security agencies or specific professionals who are likely to be at the spot. For example, in an aircraft-hijacking crisis, the first responders will be the aircraft crew including the pilot and cabin crew. The professional executors are people who are going to implement the plans made by decision-makers. These executors may be professionally trained to handle specific situations, and may have specific equipment to take care of the peculiar conditions in which they have to operate. The training needs of these people include developing quick-reaction decision-making skills under pressure, besides the special preparation needed to take care of particular conditions.

Technological Resources

The technological infrastructure and expertise required by the central organisation, first responders, decision-makers and professional executives need to be established. It may so happen that the infrastructure and expertise are not used for a long while, as crisis situations may not occur regularly. However, with them in place the organisation can only handle the crisis and disaster better than it would without them.

Documentation and Databases

It is important that proper documentation and databases should be constructed and maintained continuously; they should relate to a variety of crisis situations in the past and how each one was handled.

Generation of Plans for Crisis and Disaster Management

During this stage it is important to generate alternative plans for handling crises and disasters. These should be reviewed periodically and their efficacy evaluated.

Public Awareness

Generating public awareness related to a variety of crisis situations is quite crucial to this phase. During the actual crisis and disaster situations media management becomes important. Hence systems and people should be there to communicate to the media using established procedures.

Countering Threats

The conditions and events that can lead to crisis situations need to be countered during this period itself. This nipping-in-the-bud approach can be quite effective if a proper understanding of the events is obtained.

9.4.2 Factors Relevant to Crisis Management

Once the crisis situation actually occurs the crisis management phase starts. The goal here is to avoid disaster as far as possible. This is the most critical phase. A wrong move now will lead to disaster. The key factors contributing in this phase are ability to rapidly form a crisis management team, communication infrastructure, crisis databases, decision-making and response execution.

Rapid Formation of a Crisis Management Team

It is of the utmost importance that as soon as the crisis avoidance organisation detects a crisis it should form a crisis management team by incorporating representatives from affected and response organisations. This team is immediately given the responsibility of managing the crisis.

Communication Infrastructure

During the time of crisis, communications become pivotal. They include those between decision-makers, first responders and professional executives as well as those between these persons and the organisational or national leadership, the media and the public.

Crisis Databases

Crisis databases and documentation during the crisis itself are also important, helping in the generation of responses not only in the present crisis but also sometimes during the next crisis as well.

Decision-making

By far the most important activity in this phase is decision-making, which includes response generation, response evaluation and response choice under the constraints of limited time and limited information.

Response Execution

The response execution is the final activity that will pull back the crisis situation from degenerating into disaster. This activity should be flexible enough to respond to a rapidly developing crisis situation.

9.4.3 Factors Relevant to Disaster Management

When disaster actually strikes it leaves the existing systems that are in place in total chaos. Yet it is important to rebuild the pieces from the disaster. This involves providing immediate relief, damage assessment, rehabilitation, reconstruction, communications, documentation and logistics support.

Relief

This includes the provision of resources to search for rescuing people, medical services and lifeline construction. It has been found in the past that immediately after a disaster the people affected reunite to help each other. They join hands in rescuing their fellow beings, take the injured to hospitals, build shelters, and carry out other relevant tasks [BOOK REF]. Therefore, it becomes important to organise the relief operations in such a manner that the work being carried out by those initially affected is not hampered.

Damage Assessment

Assessing the damage done to the people, infrastructure and strategic capability of the nation, group or business is important and should be carried out with a view to responding to further rehabilitation and reconstruction.

Rehabilitation

This includes restoring public, community or business assets as well as private assets to their pre-disaster condition – usually a long-drawn-out affair.

Reconstruction

This implies improving upon the conditions prevailing before the disaster so that the possibility of disaster is minimised.

Communication

Communication structure becomes an important element during disaster management, especially during relief operations. As disaster striking at an area demolishes the infrastructure to a great extent, so if disaster relief operations are to happen efficiently, the immediate need is to set up a communication infrastructure.

Protecting the Emergency Responders

During 9–11 December 2001, a conference of emergency responders was held in New York. The emergency responders discussed their experiences after the 11 September 2001 attacks on the World Trade Center and the Pentagon. The proceedings of the conference are documented in [6]. The key findings with regard to the availability, use, performance and management of personal protective equipment (PPE) and the strategic policy issues are listed below.

Clear operating guidelines are needed and should be embedded in the training of emergency responders to use PPE efficiently. These guidelines and protocols should identify different PPE needed for different responders. The guidelines need to be easy to use and to not interfere with the performance of critical life-saving missions by the emergency responders.

The cost of a comprehensive suite of PPE for each individual emergency responder will be enormous. Hence efficient ways of delivering the relevant equipment to the specific disaster site should be implemented.

There is a considerable scope for R&D to create better and more relevant PPE. The new technologies, developed cost-effectively, can help the emergency responders to a great extent.

Equipment standardisation and interoperability is a major issue. Uniform training, maintenance and use protocols need to be formalised.

On-site safety management was another issue. The disaster site needs to be under a defined management structure, with clear lines of authority and responsibility.

Documentation

It may seem almost useless to keep on documenting events during every phase of the disaster, but it is essential to analyse the reasons, responses and results of policies and disaster management systems in place.

Logistics Support

Logistics support needs to be provided to first responders and professional executives so that they can work with maximum efficiency.

The analysis of these factors at the various stages of disaster management is the major prerequisite to formulating a strategy for disaster management. In the next section we demonstrate the use of the analytic hierarchy process (AHP) to rank these factors in terms of relative importance at each stage of disaster management.

9.5 Ranking the Factors Using Expert Judgement

At each stage of disaster management, there are certain activities that an organisation needs to perform, and their relative importance is an important input in formulating

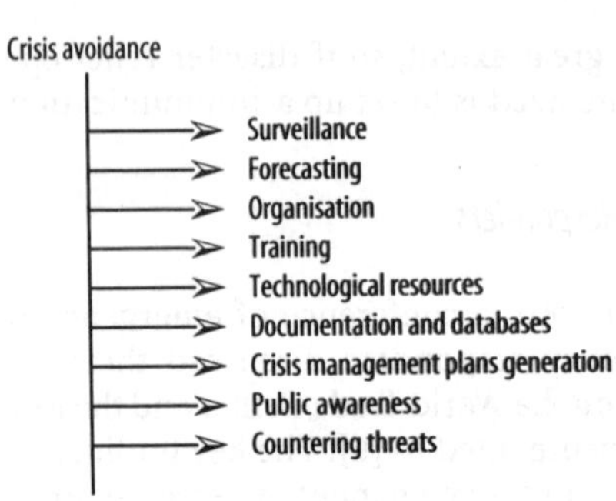

Figure 9.3 Factors affecting the crisis avoidance phase.

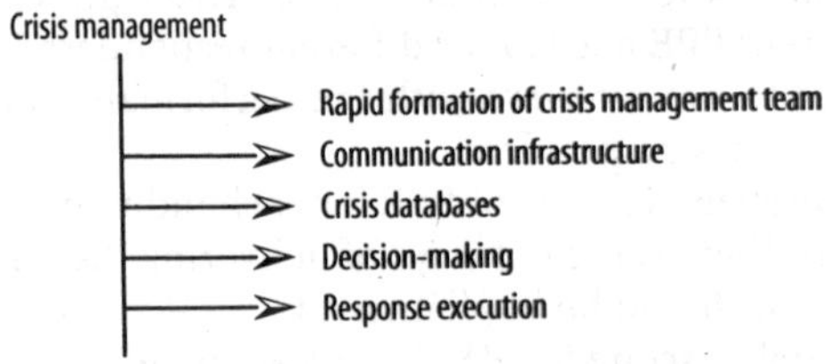

Figure 9.4 Factors affecting the crisis management phase.

Figure 9.5 Factors affecting the disaster management phase.

a strategy. The hierarchy of activities for crisis avoidance is shown in Figure 9.3, and those for crisis management and disaster management in Figures 9.4 and 9.5 respectively.

Experts were asked to rank these factors for the three phases of disaster management. The questions asked were related to the specific scenario of a weapons of mass destruction (WMD) disaster. (As noted above, WMD is an umbrella acronym for nuclear biological and chemical (NBC) weapons.) The WMD crisis can occur in an armed-conflict scenario between two state actors which escalates to an NBC war. It can occur in a threat by non-state actors (e.g. terrorist organisations) to use NBC weapons. It can also be a sudden crisis arising out of a malfunctioning nuclear power station. Although the experts were asked to rank these activities in various phases of disaster management in the specific WMD scenario, the results can be taken as a generic framework for any disaster management situation.

Table 9.1 Pairwise comparison matrix for crisis avoidance phase.

	S	*F*	*O*	*T*	*TR*	*DD*	*CMP*	*PA*	*CT*	*NEV*
Surveillance (S)	1	3	1/3	1/3	1/3	3	1/5	1/7	1/9	0.033
Forecasting (F)	1/3	1	1/5	1/7	1/7	1	1/7	1/7	1/9	0.018
Organisation (O)	3	5	1	1	1	3	1	1/3	1/9	0.077
Training (T)	3	7	1	1	1	3	1/5	1/7	1/9	0.062
Technological resources (TR)	3	7	1	1	1	5	1/3	1/5	1/7	0.071
Documentation and databases (DD)	1/3	1	1/3	1/3	1/5	1	1/5	1/7	1/7	0.023
Crisis management plan (CMP)	5	7	1	5	3	5	1	1	1	0.175
Public awareness (PA)	7	7	3	7	5	7	1	1	1	0.229
Countering threats (CT)	9	9	9	9	7	7	1	1	1	0.313
Eigenvalue = 9.848, CI = 0.106, CR = 0.07										

Table 9.2 Pairwise comparison matrix for crisis management phase.

	CMT	*C*	*CD*	*DM*	*RE*	*NEV*
Rapid formation of crisis management team (CMT)	1	1	3	1	1	0.226
Communication infrastructure (C)	1	1	3	1	1	0.226
Crisis databases (CD)	1/3	1/3	1	1/3	1/5	0.069
Decision-making (DM)	1	1	3	1	1	0.226
Response execution (RE)	1	1	5	1	1	0.253
Eigenvalue = 5.031, CI = 0.008, CR = 0.007						

Table 9.3 Pairwise comparison matrix for disaster management phase.

	Rel	*DA*	*Reh*	*Rec*	*Co*	*Do*	*LS*	*NEV*
Relief (Rel)	1	3	5	5	5	1	3	0.328
Damage assessment (DA)	1/3	1	3	3	3	1	3	0.185
Rehabilitation (Reh)	1/5	1/3	1	1/3	1	1/3	1	0.052
Reconstruction (Rec)	1/5	1/3	3	1	3	3	3	0.163
Communication (Co)	1/5	1/3	1	1/3	1	1/3	1/3	0.046
Documentation (Do)	1	1	3	1/3	3	1	3	0.158
Logistics support (LS)	1/3	1/3	1	1/3	3	1/3	1	0.069
Eigenvalue = 7.761, CI = 0.127, CR = 0.096								

The questions were put in designated formats and put to experts. The pairwise matrices formed from the inputs are shown in Tables 9.1, 9.2 and 9.3 for the three phases of crisis avoidance, crisis management and disaster management, respectively. The normalised right eigenvector (NEV) corresponding to the principal eigenvalue of these matrices gives the relative importance of the parameters in each phase. For example, in the crisis avoidance phase (see Table 9.1) the experts giving the inputs feel that countering threats is the most important activity (importance = 0.313), while public awareness (importance = 0.229) and generating crisis management plans (importance = 0.175) are also important. However, forecasting (importance = 0.018) and documentation and databases (importance = 0.023) have been given very low importance.

A total of seven experts were asked to give these inputs. The problem was to combine the inputs so that a consensus ranking could be obtained. For combining these judgements the geometric mean of the final rankings given by all the experts was used. The results are shown in Tables 9.4, 9.5 and 9.6 for the phases of crisis avoidance,

Table 9.4 Combining expert judgements for factors affecting the crisis avoidance phase.

	E1	*E2*	*E3*	*E4*	*E5*	*E6*	*E7*	*GM*	*NGM*
Surveillance	0.033	0.231	0.254	0.225	0.202	0.273	0.105	0.159	0.197
Forecasting	0.018	0.169	0.179	0.051	0.209	0.019	0.080	0.071	0.088
Organisation	0.077	0.019	0.025	0.185	0.049	0.207	0.187	0.075	0.093
Training	0.062	0.058	0.183	0.082	0.113	0.035	0.114	0.082	0.102
Technological resources	0.071	0.059	0.063	0.150	0.272	0.173	0.159	0.117	0.145
Documentation and databases	0.023	0.112	0.041	0.043	0.058	0.079	0.096	0.057	0.071
Crisis management plan	0.175	0.103	0.091	0.170	0.048	0.094	0.086	0.101	0.126
Public awareness	0.229	0.024	0.072	0.053	0.023	0.047	0.102	0.058	0.073
Countering threats	0.313	0.225	0.093	0.040	0.025	0.074	0.070	0.086	0.106
Eigenvalue	9.85	9.83	14.13	10.5	9.65	13.49	13.47		
CI	0.106	0.104	0.649	0.118	0.081	0.562	0.559		
CR	0.07	0.07	0.45	0.08	0.06	0.39	0.39		

Table 9.5 Combining expert judgements for factors affecting the crisis management phase.

	E1	*E2*	*E3*	*E4*	*E5*	*E6*	*E7*	*GM*	*NGM*
Formation of crisis management team	0.226	0.225	0.119	0.374	0.201	0.517	0.149	0.231	0.272
Communication infrastructure	0.226	0.171	0.207	0.051	0.098	0.038	0.107	0.107	0.126
Crisis databases	0.069	0.110	0.060	0.046	0.037	0.301	0.213	0.090	0.107
Decision-making	0.226	0.420	0.252	0.221	0.199	0.051	0.173	0.191	0.225
Response execution	0.253	0.075	0.362	0.308	0.465	0.092	0.358	0.228	0.269
Eigenvalue	5.030	5.505	5.400	5.140	5.195	5.856	5.768		
CI	0.008	0.126	0.100	0.036	0.049	0.214	0.192		
CR	0.007	0.113	0.089	0.032	0.043	0.191	0.172		

Table 9.6 Combining expert judgements for factors affecting the disaster management phase.

	E1	*E2*	*E3*	*E4*	*E5*	*E6*	*E7*	*GM*	*NGM*
Relief	0.328	0.358	0.161	0.351	0.250	0.366	0.480	0.313	0.339
Damage assessment	0.185	0.122	0.115	0.293	0.287	0.037	0.126	0.139	0.151
Rehabilitation	0.052	0.034	0.083	0.071	0.062	0.238	0.311	0.090	0.098
Reconstruction	0.163	0.060	0.245	0.099	0.093	0.050	0.204	0.112	0.122
Communication	0.046	0.027	0.071	0.029	0.043	0.149	0.146	0.059	0.064
Documentation	0.158	0.105	0.247	0.087	0.232	0.077	0.071	0.124	0.135
Logistics support	0.069	0.292	0.078	0.071	0.034	0.084	0.094	0.084	0.091
Eigenvalue	7.76	11.78	7.59	7.67	7.49	11.25	9.25		
CI	0.127	0.797	0.112	0.112	0.082	0.708	0.373		
CR	0.096	0.604	0.085	0.085	0.062	0.536	0.283		

crisis management and disaster management, respectively. The consistency ratio threshold was put at 0.15 rather than the traditional 0.1. This was done to incorporate the unstructured nature of the parameters. It is proposed that for very abstract parameters an inconsistency of up to 0.2 should be allowed.

The normalised geometric mean gives the consensus ranking of all the experts. However, a look at the consistency ratio shows that experts E2, E6 and E7 have CR values of 0.45, 0.39, and 0.39 respectively. These values are considerably more than the accepted value of consistency ratio, which is 0.15. Similarly in Table 9.5, experts E6 and E7 have CRs of 0.191 and 0.172, which also exceed the accepted value of 0.15. The CR of experts E2, E6 and E7 in Table 9.6, which are 0.604, 0.536 and 0.283 respectively, also all exceed 0.15.

Table 9.7 Importance of factors affecting the crisis avoidance phase after eliminating the inputs of inconsistent experts (E3, E6 and E7).

Crisis avoidance phase	*NGM (final) = F*	*NGM (without elimination from Table 9.4) = E*	*% difference 100 × (E − F)/E*
Surveillance	0.178	0.197	10.54
Forecasting	0.099	0.088	−10.73
Organisation	0.079	0.093	17.97
Training	0.099	0.102	2.77
Technological resources	0.149	0.145	−2.92
Documentation and databases	0.066	0.071	7.96
Crisis management plan	0.144	0.126	−12.42
Public awareness	0.066	0.073	9.84
Countering threats	0.120	0.106	−11.42

Table 9.8 Importance of factors affecting crisis management phase after eliminating inputs of inconsistent experts (E6 and E7).

Crisis management phase	*NGM (final) = F*	*NGM (without elimination from Table 9.5) = E*	*% difference 100 × E − F)/E*
Rapid formation of crisis management team	0.236	0.272	15.43
Communication infrastructure	0.145	0.126	−13.04
Crisis databases	0.066	0.107	62.49
Decision-making	0.279	0.225−19.26	
Response execution	0.275	0.269	−2.16

Table 9.9 Importance of factors affecting the disaster management phase after eliminating the inputs of inconsistent experts (E2, E6 and E7).

Disaster management	*NGM (final) = F*	*NGM (without elimination from Table 9.5) = E*	*% difference 100 × (E − F)/E*
Relief	0.277	0.339	18.43
Damage assessment	0.218	0.151	−44.35
Rehabilitation	0.070	0.098	28.60
Reconstruction	0.147	0.122	−20.29
Communication	0.048	0.064	25.61
Documentation	0.178	0.135	−31.52
Logistics support	0.064	0.091	30.08

The next step is to eliminate all the inconsistent experts from the geometric mean computation and recompute the NGM of the consistent experts. Let this result be called the accepted consistent normalised geometric mean (ACNGM). These values for crisis avoidance phase are shown in Table 9.7. The table also shows the percentage difference between the NGM and the ACNGM. It can be seen that the maximum variation of 17.97% is in the importance of the factor "organisation" and the minimum variation is in the importance of the factor "training".

Similar analyses done for the crisis management and disaster management phases are shown in Tables 9.8 and 9.9 respectively. It can be seen that the maximum variation of 62.49% is observed in the importance of the factor "crisis databases" in the crisis management phase (see Table 9.8). In the disaster management phase maximum variation is seen in the importance of damage assessment, which is −44.35% (see Table 9.9).

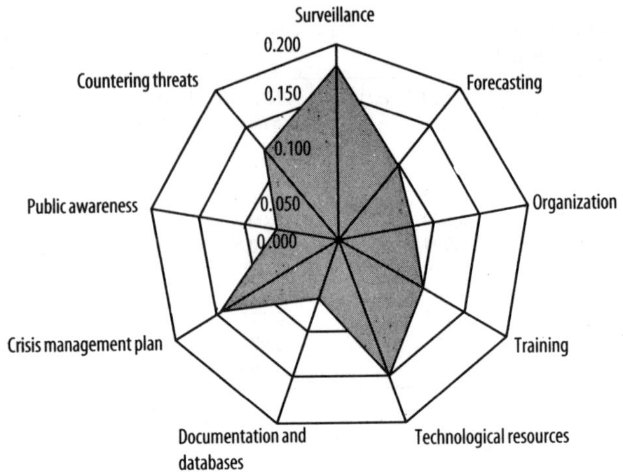

Figure 9.6 Crisis avoidance Kiviat chart.

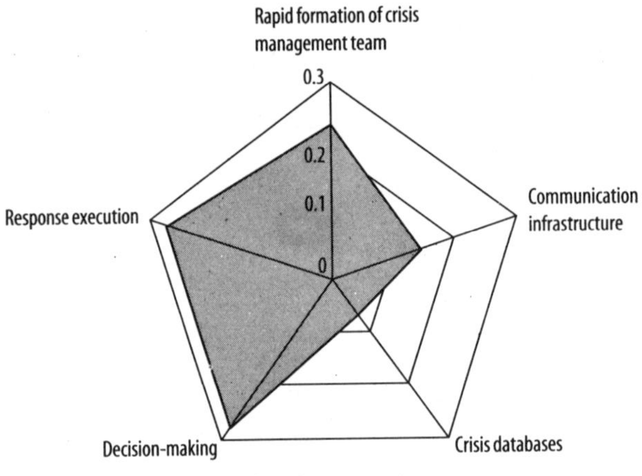

Figure 9.7 Crisis management Kiviat chart.

The Kiviat chart shown in Figure. 9.6 for the crisis avoidance phase shows that the most important factors in this phase are surveillance, technological resources, crisis management plans, and countering threats. This is an important conclusion which it is not possible to ascertain from the inputs trivially.

Figures 9.7 and 9.8 show the Kiviat charts of the relative importance of activities in the crisis management phase and the disaster management phase respectively. As can be seen, in the crisis management phase the most important activities are decision-making, response execution, and the rapid formation of crisis management teams (see Figure. 9.4). In the disaster management phase the Kiviat chart is skewed towards relief and damage assessment, with documentation and reconstruction making important contributions as well (see Figure. 9.5).

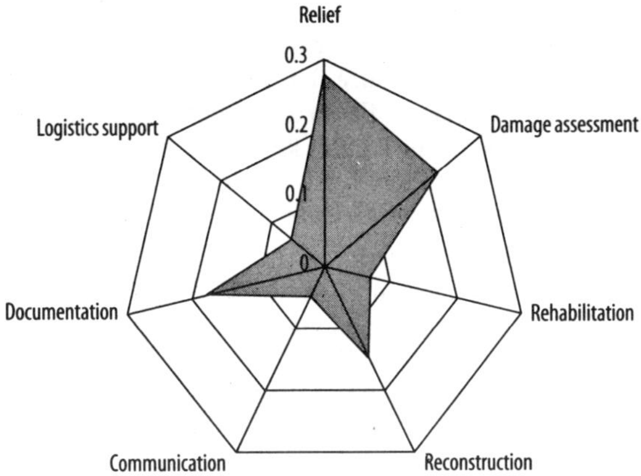

Figure 9.8 Disaster management Kiviat chart.

9.6 A Generic Strategy for Disaster Management

Is it possible to extract from the above analysis a generic strategy for disaster management? We have seen how a normal situation can develop into a crisis and then into a disaster. We have treated this transition in three distinct phases – crisis avoidance, crisis management and disaster management. We have identified key activities that are important in each of these phases and the relative importance of these activities in each phase. We can now select some key recommendations for a generic strategy for disaster management as follows:

- A centralised unit is needed for disaster management of all kinds. This central disaster management division (CDMD) could be a small team of techno-business experts in a business setting or it could be a separate ministerial level organisation in the central government of a nation.
- This division should have predefined, reliable and unambiguous communication links and communication protocols with top leadership in all the relevant divisions of the organisation. The central disaster management division will be the nodal agency for surveillance and monitoring, generating crisis management plans, and creating or incorporating technological solutions and capabilities in the disaster management and crisis management infrastructure. This division should produce plans to evaluate threats that can lead to disasters so that these can be responded to in the crisis avoidance phase. The division will also look after the training needs of decision-makers, early responders and emergency responders. In addition it will create programmes for public awareness.
- As soon as a crisis situation arises, a crisis management team (CMT) will be formed by the central disaster management division to respond to the crisis. This team will be formed with decision-makers from all relevant units, who will be completely devoted to solving the crisis and to making sure that it does not go to a disaster situation. However, if it is likely to result in disaster, or even otherwise, then the CMT will immediately prepare the emergency response teams with the help of the CDMD. The CMT will be formed with people who

have training in crisis management or have managed crises in the past. The CMT will be tasked to prepare responses and create alternatives to solve the crisis with the least possible loss for the organisation or government. The CDMD along with the CMT will make sure that all possible resources are provided to execute the chosen response to the crisis situation. As has been shown in the analysis, response execution is the key. The execution unit, e.g. military commandos rescuing passengers from a hijacked aircraft, need to be properly trained to execute the chosen response and should have clear, unambiguous instructions to this end.

- In the event of a disaster happening, the CMT and CDMD will mobilise all the resources, or if the crisis management phase is long enough the CDMD will form a disaster management team (DMT) which will have some members of the CMT but more members from relief-providing organisations. The DMT will initiate relief operations as soon as possible and will also with the help of all organisations start the process of damage assessment. Immediate relief to the affected will be the topmost priority of the DMT. Detailed documentation of the disaster will be made and kept in the crisis and disaster databases. These will be used by the CDMD to start the process of rehabilitation and reconstruction. Since rehabilitation is usually a long-drawn-out affair, the CDMD along with other groups will come out with action plans for rehabilitation, assess the likely costs and also give recommendations to the government or business management on the likely action plan.

References

1. Denis H (1996) Natural and technological disaster management: Uncertainty in technical, socio-political and scientific issues. In: El-sabh MI et al. (eds) Land-based and Marine Hazards. Kluwer.
2. Oakley DJ (1993) Keynote paper: A national disaster preparedness service. In: Natural Disasters. Telford, London.
3. Smallman C and Weir D (1999) Communication and cultural distortion during crisis. Disaster Prevention and Management 8(1).
4. Payne CF (1999) Contingency plan exercises. Disaster Prevention and Management 8(2).
5. Blanchard JC and Davis LM (2002) Are Local Health Responders Ready for Biological and Chemical Terrorism? Issue Paper IP-221-OSD, Rand Corporation, Santa Monica, CA.
6. Jackson B et al. (2002) Protecting Emergency Responders – Lessons Learned from Terrorist Attacks. CF-176-OSTP. Rand Corporation, Santa Monica, CA.
7. Parachini JV, Davis LE and Liston T (2003) Homeland Security, white paper WP127. Rand Corporation, Santa Monica, CA.

10 Weapon Systems Acquisition for Defense Forces

10.1 Introduction

Technology is changing the shape of warfare through multidimensional influences on the components of war-making machinery. Weapon system acquisitions by a country's armed forces require a critical evaluation of the requirements and capabilities of the alternative systems available, the adversary's equipment and the cost. Also, owing to the longer life-period of weapon systems in modern armies, changing demands on the armed forces, a dynamic world situation and the rapid pace of technological change, it becomes imperative to make decisions related to major weapon system acquisitions in an organised and comprehensive manner. Before inducting any weapon system, an army should critically evaluate the available alternatives.

Static comparison of the various features of a weapon system with the corresponding features of an alternative system in order to evaluate its battle performance has been carried out using well-established methodologies. The basis of these methodologies has been computation or assignment of a numerical value to a weapon system, indicating the estimate of its lethal capability. This numerical value has been called the fire power score (FPS; see Appendix 10.1 for a glossary of acronyms used in this chapter). The FPS methodology has been extended in Bhushan and Jain [1] to incorporate the survivability aspect of the weapon system along with the lethality of the system into a new index called the weapon power score (WPS). In this chapter, WPS methodology is extended to incorporate the operability and integrational aspects of the weapon system into the WPS. Further, the WPS methodology is generalised to incorporate the effects of various terrains and the nature of future battles in which the weapon system will operate. This new index is called the generalised weapon power score (GWPS). The application of the GWPS is shown below (Section 7.3) in the comparative rating of the performance of main battle tanks.

10.2 Fire Power Scores (FPS) and Weapon Power Scores (WPS)

The basic idea in fire power scores (FPS) methodologies is to assign a numerical value to different weapons indicating their war-making capability. Various firepower score methodologies have been developed on the basis of expert judgement, historical data analysis and combat simulation. Some of these methodologies are: the weapon effectiveness index (WEI)/weapon unit value (WUV), based on expert judgement [2]; the potential anti-potential (PAP) method, which uses combat simulations [3]; and the operational lethality index (OLI) based on historical data analysis [4]; etc.

WEI/WUV have been developed based on expert judgement obtained using the Delphi method. Relative weights are assigned to attributes of the weapon system from the point of view of combat effectiveness. Experts for each weapon system rate its attributes in the specified category. Instead of the Delphi method Lee and Ahn [5] have used the AHP in the WEI/WUV method to assess the relative effectiveness of various weapons of the Korean army. The PAP method, also called the eigenvalue method, computes firepower scores based on combat simulations using a high-resolution combat model. From the combat simulation, killer-victim scoreboards giving the attrition of a weapon category by the weapons of the enemy side are generated. These scores are arranged in killer-victim matrices, which are used to generate the firepower scores. The OLI method is part of the quantified judgement method of analysis (QJMA), which is used for combat evaluation. However, for the purpose of this book, OLI has been considered as the fire power score.

Measuring Military Power

Who is more powerful militarily is something that everyone – be they a general, a politician, a business person or just the man in the street – always wants to know. This interest is natural, as war is a situation that is extraordinary. It bears upon the very core of human existence. All of us are affected by war, since wars are something that seem to occur regularly, despite the centuries of so-called civilised societies. People have used various methods to account for the raw military power of the adversaries in order to gauge the war outcomes before the actual events. Although victory or loss has depended historically on the way the weapons have been used in the specific operational situations, i.e. the tactics, and the way the whole of the operations has been planned, i.e. the strategy, raw weapon power has always been an important input for war planners.

This measurement becomes all the more important at the strategic decision-making level in war or conflict situations. If the raw combat power of an adversary comes out to be less than a third of your own then the chances of you winning the war are very high. This 3:1 golden rule, although contested by many, has been the guiding principle for strategy, operations and tactical attacks. However, measuring combat power based on the number of the weapon systems alone leads to highly inaccurate estimates. This quantity, although a significant contributor, can be nullified to a great extent by quality and superior doctrine. The quality has been measured by raw lethal capability for centuries, but now technology is changing the whole face of warfare. Now it is as important to invest in systems that help in integrating the fighting units in a network, weapon systems that can operate in extreme weather conditions and at night, and systems that have enough on-board protection to preserve themselves in the more lethal and more dangerous, dynamic battlefields of today.

Most of the FPS methodologies for so-called static analyses [6] give less importance to other factors such as the self-protection capability of a weapon system, its ability to operate in all weather conditions and at nighttime etc. An attempt has

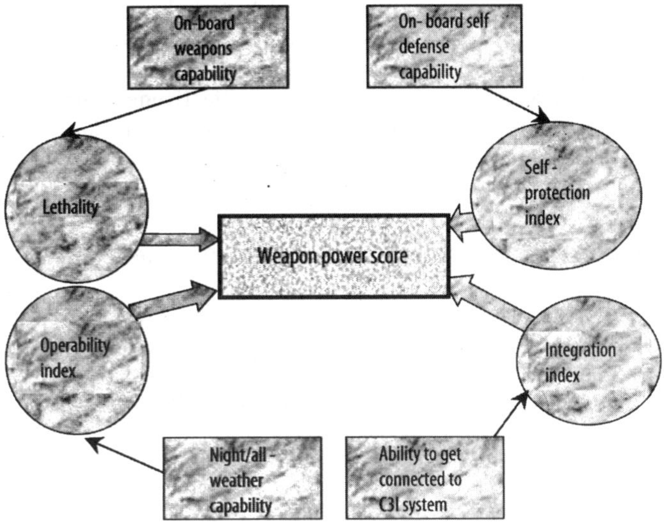

Figure 10.1 Weapon power score.

been made to augment FPS with the survivability of the weapon system to give a realistic picture of the combat potential of a force [1].The new score is called the weapon power score (WPS). In this book, weapon power scores are modified and extended to incorporate other factors such as on-board self-protection capability, operability and ability to communicate with other weapon systems through command, control, communications, and intelligence (C3I) links. Expert judgement has been used for the evaluation of WPS. The evaluation of WPS involves assigning a numerical value to each weapon system of an armed force indicating its combat effectiveness. WPS is defined as follows(see Figure 10.1):

$$\text{WPS} = \text{Operational lethality index (OLI)} \times (1 + \text{Self-protection index (SPI)}) \times (1 + \text{Operability index (OI)}) \times (1 + \text{Integration index (II)}) \quad (10.1)$$

The lethality of a weapon is expressed as the operational lethality index (OLI) as defined by Dupuy [4]. In this method, empirical formulae based on historical combat data are used for determining these indices. The method divides all weapon systems into two broad categories, namely mobile fighting machines and non-mobile fighting machines. For non-mobile weapons the OLI is defined as a function of various factors such as rate of fire, number of potential targets per strike, range factor etc. For mobile fighting machines, the OLI is computed, adding the separately calculated OLIs of all the weapons on the mobile fighting machine, and multiplying this result by several performance factors.

10.3 Weapon Power Scores for Main Battle Tanks

An important element of a modern army continues to be the armoured force, and main battle tanks (MBT) form the major component of armour. The evaluation procedure for WPS of MBTs is described in this section. The WPS is composed of four performance indices as described above. The lethality is reflected in the OLI of the tank. The parameters used in OLI computation of MBTs include rate of fire, range, accuracy, reliability of on-board weapons, mobility-related factors and certain armour-related factors such as fire control effectiveness, ammunition supply effect etc. Empirical formulas for computing these factors have been provided based on historical data.

10.3.1 Self-protection, Operability and Integration Indices

Besides lethality, it is observed that other factors such as on-board self-protection, operability and capability to get integrated with the C3I system also play important roles in weapon effectiveness. Weapon capability is enhanced if the weapon has characteristics that improve its ability to survive in the battle. The characteristics may include on-board radar, decoys/chaff, electronic counter measures (ECM), electronic support measures (ESM), armour protection etc. There are certain on-board systems that enhance the self-protection capability of the weapon system. This is reflected in the self-protection index (SPI), with a value in the range 0 to 1. The value of SPI for various weapon systems also takes into account the environment in which the weapon system will be operating. The effectiveness of a weapon system also depends upon its ability to operate in adverse weather and/or environmental conditions and night operations. It is reflected in the operability index (OI) of the weapon system. The OI indicates the ability of the weapon system to operate in adverse weather and night operations. The value of OI is chosen from the interval 0 to 1. Another factor enhancing the effectiveness of a weapon system is its ability to become integrated in the C3I system. This factor depends upon the communication links of the weapon system with various surveillance systems, and with command and control systems. It is represented as the integration index (II) of the weapon system. The value of II is also chosen from the interval 0 to 1. The factors to be considered for the evaluation of the self-protection index, the operability index and the integration index of MBTs are given in Table 10.1.

Expert judgement has been used for grading various tanks with respect to features contributing to the self-protection, operability and integration indices. Experts are also asked to assign weights to features contributing to these indices. The summation of the products of the weights and the gradings assigned to the various features/components contributing to self-protection, operability and integration gives the SPI, OI and II of the tank. The grading is assigned as per the guidelines given in Table 10.2.

Table 10.1 Factors to be considered for the evaluation of SPI, OI and II of MBTs.

II	*SPI*	*OI*
Communication links with other armoured vehicles and UAV/AOP	Built-in-armour, reactive armour, top attack protection NBC protection, fire and explosion suppression, silhouette, agility	Night vision, fording, amphibious capability, ability to get air-dropped

Table 10.2 Grading scale for weapon capability related to various attributes.

Capability	*Grading*
No capability	0.0
Marginal Capability	0.33
Adequate Capability	0.67
Substantial Capability	1.00

Table 10.3 Format for evaluating integration indices.

Communications capability	*Weight* (W_i)	*Capability parameter*					*Mean capability* (C_i)	*Remarks*
		Supported by system	*Security*	*Bandwidth*	*Addressing mode*	*Range*		
Voice	$W_1 = 0.3$	C_{11}	C_{12}	C_{13}	C_{14}	C_{15}	$(\Sigma_j C_{1j})/5$	
Data	$W_2 = 0.1$	C_{21}	C_{22}	C_{23}	C_{24}	C_{25}	$(\Sigma_j C_{2j})/5$	
Image transmission	$W_3 = 0.1$	–	–	–	–	–	–	Image processing and transmission
Map display and processing	$W_4 = 0.1$	–	–	–	–	–	–	
Decision support system	$W_5 = 0.1$	–	–	–	–	–	–	
Data transmission	$W_6 = 0.1$	–	–	–	–	–	–	Integration with UAV
Image	$W_7 = 0.1$	–	–	–	–	–	–	
Map displays	$W_8 = 0.1$	C_{81}	C_{82}	C_{83}	C_{84}	C_{85}	$(\Sigma_j C_{8j})/5$	

Table 10.3 gives the procedure for evaluating the integration index of the MBTs. Values given to weights (W_i) are hypothetical ones assumed for the purpose of illustration. For real evaluation, these inputs should be obtained from experts.

The integration index is computed as follows:

$$\Pi = \sum_{i=1}^{8} W_i C_i \tag{10.2}$$

where $\Sigma_{i=1}^{8} W_i = 1.0$. The values of the capabilities (C_{ij}) are assigned as per the grading given in Table 10.2. A similar procedure is used to evaluate the SPI and OI of the MBTs.

Example

Let us suppose that an army requires the evaluation of three MBTs with characteristics as shown in Table 10.4. Let the costs of the three tanks be in the ratio 1:2:3, i.e. T1 costs say 3 million units, T2 costs 6 million and T3 costs 9 million. The weapon power scores can be utilised for the cost-effectiveness evaluation of the tanks.

Table 10.4 Characteristics of hypothetical tanks.

Characteristic		*Tank T1*	*Tank T2*	*Tank T3*
Lethality				
Main gun		105 mm	125 mm with anti-tank missile firing capability	120 mm
Secondary armament		12.7 mm AAG and 7.62 mm machine-gun	12.7 mm AAG and 7.62 mm machine-gun	12.7 mm AAG and 7.62 mm machine-gun
Fire control system		Primitive	Automated fire control system	Automated fire control system
Ammunition		HE, APFSDS (steel core), HEAT	HE, APFSDS (soft core), HEAT	HE, APFSDS (soft core), HEAT
Weight	a_i	*Self-protection rating (g_I) for each tank*		
Built-in armour	0.25	Adequate (0.67)	Adequate (0.67)	Substantial (1.0)
Add-on Reactive armour	0.1	No (0.0)	Adequate (0.67)	Substantial (1.0)
Top attack protection	0.1	No (0.0)	No (0.0)	Marginal (0.33)
Active armour	0.1	No (0.0)	Adequate (0.67)	No (0.0)
NBC protection	0.1	Marginal (0.33)	Adequate (0.67)	Adequate (0.67)
Fire and explosion suppression system	0.05	Adequate (0.67)	Substantial (1.0)	Substantial (1.0)
Low silhouette	0.15	Substantial (1.0)	Substantial (1.0)	Marginal (0.33)
Agility	0.15	Adequate (0.67)	Adequate (0.67)	Substantial (1.0)
Weight	b_i	*Operability rating (h_I)for each Tank*		
Night vision	0.5	Marginal (0.33)	Adequate (0.67)	Substantial (1.0)
Amphibious capability	0.2	Substantial (1.0)	Marginal (0.33)	Marginal (0.33)
Air mobility	0.2	Adequate (0.67)	No (0.0)	No (0.0)
Fording	0.1	Adequate (0.67)	Adequate (0.67)	Substantial (1.0)

Values given in parentheses are the gradings of the tank parameters as per Table 10.2.

The SPI and OI of the three tanks are computed as follows:

$$SPI = \sum_{i=1}^{8} a_i g_i \tag{10.3}$$

$$OI = \sum_{i=1}^{4} b_i h_i \tag{10.4}$$

where each a_i is a weight assigned to a parameter contributing to SPI such that $\Sigma_{i=1}^{8} a_i = 1.0$ and each bi is a weight assigned to OI parameters such that $\Sigma_{i=1}^{4} b_i = 1.0$. Also, g_i and h_i are capability gradings of the tank based on the scale given in Table 10.2.

The parameter evaluation of the integration indices of three tanks is given in Table 10.5.

Let the basic OLI of each of the three tanks come as shown in Table 10.5. The SPI, OI and II of each tank as computed using the procedure described above are also shown in Table 10.6.

Table 10.5 Evaluation of integration indices for the three tanks.

Communication capability	W_I	*Capability parameters*						
		Tank	*Supported by the system*	*Security*	*Band-width*	*Addressing modes*	*Range*	*Mean capability*
Voice	0.3	T1	0.67	0.67	0.67	0.33	0.67	0.602
		T2	1.00	0.67	0.67	0.67	0.67	0.736
		T3	1.00	1.00	0.67	0.67	1.00	0.868
Data	0.1	T1	0.0	0.0	0.0	0.0	0.0	0.0
		T2	0.67	0.67	0.33	0.33	0.67	0.534
		T3	1.00	1.00	0.67	0.67	0.67	0.802
Image processing and transmission								
Image transmission	0.1	T1	0.0	0.0	0.0	0.0	0.0	0.0
		T2	0.67	0.67	0.33	0.33	0.67	0.534
		T3	1.00	1.00	0.67	0.67	0.67	0.802
Map display and processing	0.1	T1	0.0	0.0	0.0	0.0	0.0	0.0
		T2	0.33	0.67	0.33	0.33	0.33	0.398
		T3	1.00	1.00	0.67	0.67	0.67	0.802
Decision support system	0.1	T1	0.0	0.0	0.0	0.0	0.0	0.0
		T2	0.0	0.0	0.0	0.0	0.0	0.0
		T3	0.67	0.67	0.33	0.33	0.33	0.466
Integration with UAV								
Data transmission	0.1	T1	0.0	0.0	0.0	0.0	0.0	0.0
		T2	0.33	0.33	0.67	0.33	0.33	0.398
		T3	0.67	0.67	0.33	0.33	0.33	0.466
Image	0.1	T1	0.0	0.0	0.0	0.0	0.0	0.0
		T2	0.33	0.33	0.67	0.33	0.33	0.398
		T3	0.67	0.67	0.33	0.33	0.33	0.466
Map displays	0.1	T1	0.0	0.0	0.0	0.0	0.0	0.0
		T2	0.33	0.33	0.67	0.33	0.33	0.398
		T3	0.67	0.67	0.33	0.33	0.33	0.466

Table 10.6 OLI, SPI, OI and II of tanks.

Tank	*OLI*	*SPI*	*OI*	*II*	*WPS*
T1	600	0.48	0.57	0.18	1 645
T2	970	0.67	0.47	0.49	3 548
T3	1 050	0.70	0.67	0.69	5 038

The weapon power score (WPS) computed using the above method does not take into account the operational environment and likely threat scenarios that the MBTs may have to face in future. In the absence of any information regarding operational conditions the WPS can be considered as the performance rating. However, in the specific operational conditions the WPS values may change owing to the effects of operational conditions on the lethality, SPI, OI and II of the MBTs. To incorporate the operational conditions the generalised weapon power scores (GWPS) are defined below.

10.4 Generalised Weapon Power Scores

In making a decision regarding the selection of an MBT for the army, a comparison of the various alternative MBTs should be made, keeping in mind the likely nature of future conflicts and the features of the terrain in which the MBT will be operating. In this regard the intensity of conflict is considered at three levels, i.e. low-intensity conflict (LIC), mid-intensity conflict (MIC) and the high-intensity conflict (HIC). The LIC is the most prevalent form of conflict facing the world today. It is defined as the conflict between the regular army and irregular troops such as terrorists, insurgents, etc. The MIC is defined as the conventional armed conflict between regular forces of two nations. HIC is defined as a conflict involving weapons of mass destruction (WMD), i.e. nuclear biological and chemical (NBC) weapons. The utility of MBTs in each of the three levels of conflict varies. Further, the terrain features in which these conflicts may occur, i.e. high mobility and a less cluttered terrain, medium mobility and a medium cluttered terrain, and low mobility and a heavily cluttered terrain, should also be reflected in the performance evaluation of the MBTs.

The well-established methodology of the analytic hierarchy process (AHP) is used to evaluate the GWPS of the various tanks. The hierarchy is shown in Figure 10.2. Experts are asked to state the relative future likelihood (l_i) of the nation facing an LIC, a conventional military operation and an NBC operation. Also obtained is the likelihood (T_{ij}) of the conflict type (i) to be in a specific terrain (j), i.e. high mobility and a less cluttered terrain, medium mobility and a medium cluttered terrain, and low mobility and a heavily cluttered terrain. These likelihood values for various conflicts are country-dependent. For example, a small country may not perceive any NBC operations or even any conventional armoured operations in its likely conflicts. However, a relatively bigger country facing hostile, conventionally armed neighbours may face more likelihood of conventional operations. Finally, the impact of a specific terrain (j) and conflict type (i) combination on the lethality, self-protection, operability and integration indices of a tank is obtained from the experts. The methodology is described below.

Let the experts give the likelihood of the army facing an NBC operation, a conventional military operation and a low-intensity conflict as l_1, l_2 and l_3 respectively, such that $\Sigma l_i = 1.0$. These numbers are obtained using the well-establishe methodology of the AHP [11]. Similarly obtained from the experts is the likelihood (T_{ij}) of specific conflict in each type of terrain, i.e. terrain with high mobility and less

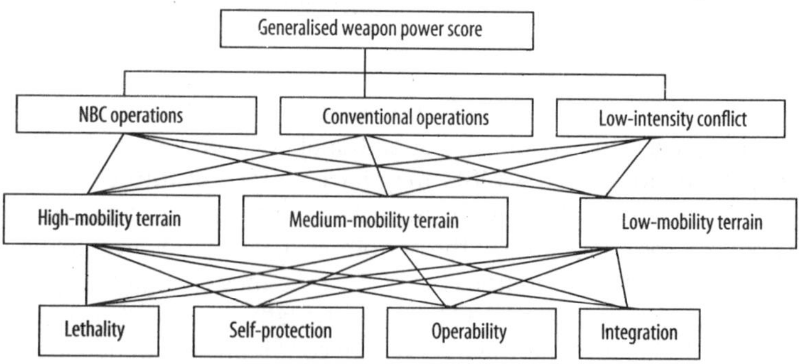

Figure 10.2 Generalised weapon power score.

Modeling Combat Dynamics

Can mathematics help in evaluating combat outcomes? Researchers in the earlier years of the twentieth century thought so. Within the first fifteen years of the last century independent researchers in the USA, the UK and Russia came out with mathematical explanations of combat dynamics. These approaches were based on differential equations. Owing to their simplicity and logic, the Lanchester models (after their inventor F.W. Lanchester) provide an attractive basis for analysing combat [7].

The main hypothesis of Lanchester and others similar [8–10] is that ancient warfare was more straightforward, consisting mostly of one-on-one duels between combatants fighting with immediate range-weapons such as swords etc. Lanchester assumed that in this scenario the casualty exchange ratio is independent of the number of combatants [7]. This seems logical, as individuals fight and destroy enemy individuals. These exchanges lead to casualties on both sides. The individual combatant capability to wield his weapon is the key factor contributing to the casualty exchange ratio. This resulted in what is called??????

Lanchester [7] proposed that in modern warfare many combatants can fire on a single target and one combatant can destroy many targets. In a particular unit of time, destruction of the number of enemy troops by each soldier may be asymmetrical. These attrition rate coefficients were the foundation parameters of the equations for modern warfare. The equations are called the Lanchester square law.

Another viewpoint developed in the combat evaluation scenarios was the assigning of numerical values to each weapon, indicating its firepower. This was the basis of current firepower score methodologies. Mathematical models have been developed which combine the two approaches to combat evaluation. In this combined model, instead of using the number of troops on both sides, combat power is used. For a detailed description of these models the reader is referred to Jaiswal [6]. Combat power is defined as the summation over all weapon types of the product of the numerical score and the quantity of weapons of each type. However, pure lethality-based weapon power is a limited indicator of weapon capability in the changed hi-tech war scenarios. Hence we propose here the new concept of weapon power scores.

clutter, terrain with medium mobility and medium clutter and terrain with low mobility and maximum clutter. Let these be as shown in Table 10.7.

Next the experts are asked to evaluate the lethality, SPI, OI and II coefficients (X, Y, Z, U) in specific conflict-type–terrain combinations, for MBT. Let these multipliers be as shown in Table 10.8.

These coefficients are estimated considering the past experience in the specific terrain, e.g. in the specific high-mobility terrain the lethality of the tank is completely utilized, hence the coefficient is 1.0. However, in the medium-mobility terrain, owing to clutter and reduced line of sight (LOS), the lethality utilization is reduced by 10%, i.e. the coefficient is 0.9. Similarly, in low-mobility terrain the lethality utilization is further reduced by 10%, i.e. the coefficient becomes 0.8. The impact of conflict type is also reflected in these coefficients. In NBC and conventional war operations the lethality of tanks is considered to be fully utilized, hence the coefficient is 1.0. In LIC the lethality is reduced by 20%, owing to the non-availability of heavy targets for which the tank

Table 10.7 Likelihood of various conflicts.

Type of conflict	*Likelihood (l_I)*	*Likelihood of conflict type (i) in terrain (j)*		
		High-mobility terrain	*Medium-mobility terrain*	*Low-mobility terrain*
Nuclear operations	0.1	$T_{11} = 0.5$	$T_{12} = 0.5$	$T_{13} = 0.0$
Conventional operations	0.3	$T_{21} = 0.4$	$T_{22} = 0.4$	$T_{23} = 0.2$
Low-intensity conflict	0.6	$T_{31} = 0.1$	$T_{32} = 0.2$	$T_{33} = 0.7$

Table 10.8 Lethality, self-protection, operability and integration multipliers of MBTs in conflict type (i) and terrain (j).

Type of conflict (i)	*Terrain (j)*	*Lethality*	*Self-protection*	*Operability*	*Integration*
NBC operations	High-mobility (H)	$X_{11} = 1.0$	$Y_{11} = 0.5$	$Z_{11} = 0.6$	$U_{11} = 0.6$
	Medium-mobility (M)	$X_{12} = 0.9$	$Y_{12} = 0.6$	$Z_{12} = 0.5$	$U_{12} = 0.5$
	Low-mobility (L)	$X_{13} = 0.8$	$Y_{13} = 0.7$	$Z_{13} = 0.4$	$U_{13} = 0.4$
Conventional operations	(H)	$X_{21} = 1.0$	$Y_{21} = 0.8$	$Z_{21} = 1.0$	$U_{21} = 1.0$
	(M)	$X_{22} = 0.9$	$Y_{22} = 0.9$	$Z_{22} = 0.9$	$U_{22} = 0.9$
	(L)	$X_{23} = 0.8$	$Y_{23} = 1.0$	$Z_{23} = 0.7$	$U_{23} = 0.7$
Low-intensity conflict	(H)	$X_{31} = 0.8$	$Y_{31} = 1.0$	$Z_{31} = 1.0$	$U_{31} = 1.0$
	(M)	$X_{32} = 0.7$	$Y_{32} = 1.2$	$Z_{32} = 0.9$	$U_{32} = 0.9$
	(L)	$X_{33} = 0.6$	$Y_{33} = 1.3$	$Z_{33} = 0.7$	$U_{33} = 0.7$

gun has been designed. The combined effects of LIC conflict and different terrain result in the lethality coefficients being 0.8, 0.7 and 0.6, in high-mobility, medium-mobility and low-mobility terrain respectively. Column 3 in Table 10.8 gives the lethality coefficients. Similar logic is used for evaluating the SPI, OI and II coefficients.

Combining the lethality, SPI, OI and II multipliers given in Table 10.8 and the likelihood of a specific conflict type occurring in specific terrain type given in Table 10.7, the effectiveness coefficients for lethality, SPI, OI and II are obtained as follows:

$$\alpha = \Sigma_i l_i \Sigma_j T_{ij} I \tag{10.5}$$

$$\beta = \Sigma_i l_i \Sigma_j T_{ij} Y_{ij} \tag{10.6}$$

$$\gamma = \Sigma_i l_i \Sigma_j T_{ij} Z_{ij} \tag{10.7}$$

$$\delta = \Sigma_i l_i \Sigma_j T_{ij} U_{ij} \tag{10.8}$$

These effectiveness coefficients are given in Table 10.9.

The generalised weapon power score (GWPS) of the MBT is given by

$$GWPS = (\alpha \times OLI) \times (1 + \beta \times SPI) \times (1 + \gamma \times OI) \times (1 + \delta \times II) \tag{10.9}$$

For all the alternative MBTs, GWPS are evaluated and used to compare the MBTs. This results in a comprehensive, more realistic, static comparison of MBTs compared with the earlier such scores. Taking into account the terrain and type of conflict, the generalised weapon power scores of the tanks come out as shown in Table 10.10.

The tank with the lowest cost/GWPS value should be chosen by the country. In this case it comes out to be tank T2 with a cost/GWPS value of 2766.

Table 10.9 Effectiveness coefficients considering conflict type and terrain for lethality, SPI, OI and II of MBTs.

Capability	*Effectiveness coefficient*	*Effectiveness coefficient value*
Lethality	$\alpha = \Sigma_t \lambda_t \Sigma_\varphi T_{t\varphi} X_{t\varphi}$	0.73
Self-protection index	$\beta = \Sigma_t \lambda_t \Sigma_\varphi T_{t\varphi} Y_{t\varphi}$	1.07
Operability index	$\gamma = \Sigma_t \lambda_t \Sigma_\varphi T_{t\varphi} Z_{t\varphi}$	0.71
Integration index	$\delta = \Sigma_t \lambda_t \Sigma_\varphi T_{t\varphi} U_{t\varphi}$	0.71

Table 10.10 GWPS of MBTs.

Tank	*Generalised weapon power score (GWPS)*	*Cost (C)*	*Cost effectiveness = cost/GWPS*
T1	1048	3 000 000	2863
T2	2169	6 000 000	2766
T3	2923	9 000 000	3079

10.4.1 Applications of Generalised Weapon Power Scores

The weapon power score describing the effectiveness of weapons as a numerical index provides a realistic weapon effectiveness measure and can be used to answer a number of questions regarding the relative performance of weapon systems. The main usage of GWPS is in the computation of the force potential of a force. For the dynamic analysis of operational battles, GWPS is used as an input to a dynamic mathematical model such as the adaptive dynamic model [12] based on difference equations giving the remaining strength of a force after the completion of an operation or combat in unit time period. A methodology combining the C4ISR system evaluation methodology such as in Juneja [13] with the GWPS described here can lead to a comprehensive approach to evaluate the force potential of an army. This can give a quantitative measure to evaluate the combat power potential of a modern network-centric armed force. The next chapter defines the comprehensive methodology.

References

1. Bhushan N and Jain RK (1998) Firepower scores and survivability: weapon power scores, GS009. In: Proceedings of Ballistics '98, 17th International Symposium on Ballistics, 23–27 March, South Africa.
2. Miller GS (1989) The Battalion Decision Aid: A Kernel Decision Support System, AFIT/GST/ ENS/89 J-4. Air Force Institute of Technology, Wright-Patterson AFB, OH.
3. Anderson LB and Miercort FA (1995) On weapons score and force strengths. Naval Research Logistics 42: 375–395.
4. Dupuy TN (1985) Numbers, Prediction and War. Hero Books, Fairfax, VA.
5. Lee, Young-Woo and Ahn, Byong-Hun (1991) Static valuation of combat force potential by the analytic hierarchy process. IEEE Transactions on Engineering Management 38(3): 237–243.
6. Jaiswal NK (1997) Military Operations Research: Quantitative Decision. Kluwer, Dordrecht.
7. Lanchester FW (1915) The principle of concentration. Engineering, October.
8. Fisk B (1905) American Naval Policy. US Naval Institute 113: 1–80.
9. Fisk (1916).
10. Osipov (1915).
11. Saaty TL (1980) The Analytic Hierarchy Process. McGraw Hill, New York.

12. Epstein JM (1990) Conventional Force Reductions – A Dynamic Assessment. Brookings Institution, Washington, DC.
13. Juneja HS, Bhushan N, et al. (1998) A methodology for evaluation of C4ISR. In: Proceedings of the IIIrd European Electronic Battlespace (Land) Symposium, 15–17 September, Cranfield University, RMCS, Shrivenham.

Appendix 10.1 Glossary

AAG	antiaircraft gun(s)
AOP	air observation post
APFSDS	armour-piercing, fin-stabilized discarding sabot
C3I	command control communications and intelligence
C4ISR	command control communication computers intelligence surveillance and reconnaissance
ECM	electronic control measure
ESM	electronic support measure
FPS	fire power score
GWPS	generalised weapon power score
HE	high-energy
HEAT	high-energy antitank
HIC	high-intensity conflict
II	integration index
LIC	low-intensity conflict
LOS	line of sight
MIC	mid-intensity conflict
NBC	nuclear biological chemical
OI	operability index
OLI	operational lethality index
PAP	potential antipotential method
QJMA	quantified judgement method of analysis
SPI	self-protection index
UAV	unmanned aerial vehicle
WEI	weapon effectiveness index
WMD	weapons of mass destruction
WPS	weapon power score
WUV	weapon unit value

11 Evaluating the Revolution in Military Affairs (RMA) Index of Armed Forces

11.1 The Revolution in Military Affairs (RMA)

A RMA has been defined to take place when

> *one of the participants in a conflict incorporates new technology, organisation and doctrine, to the extent that victory is attained in the immediate instance, but more importantly, that any other actors who might wish to deal with that participant or activity must match, or counter the new combination of technology, organisation, and doctrine in order to prevail.*

Looking at the current RMA, it is difficult to describe one single factor that has created it. Is it the revolutionary *technology* that is driving the change, is it the revolutionary *adaptation* by the military organisations or it is the *impact* of geopolitical and technological changes that is the constituent of the RMA? There are different opinions. It is said that dynamic global requirements, shifting priorities and a world in flux are bound to create such discontinuities. The second view is that the evolving technology, weapons, military organisations and doctrine are creating this change together. This is the dominant view. However, a major group of analysts consider that a true RMA is unlikely but rather there will be continuous evolution in equipment, organisations and tactics to adjust to changes in technology and the international environment. We assume that all these three conceptions are valid and that it is their interplay which has created the revolution in military affairs.

Integration of complex systems as an effective force is the key to the wars of the future. The old structures of independent operations of logistics, holding defences, mobile strikes, air force, navy, intelligence gathering, etc. need to be combined together through a series of innovative intellectual inputs. This integration of diverse systems is needed to support combat operations. The lethal system comprising various weapons, weapon systems, fighting troops, military hardware etc. is only one part of the modern combat. The other part, reflected in integration and support such as command and control (C2), communications, information warfare (IW), intelligence, surveillance and reconnaissance (ISR) etc., has become so important that today's armed forces can't think of fighting in modern battlefields without the proper combat support systems. Recently there have been numerous studies related to combat support system evaluation based on simulation and other analytical methods [3–6].

Here we propose a methodology for the evaluation of the effectiveness of the integration and support systems, which create the RMA capabilities, in a force compared with other forces. The evaluation is based on finding the various interdependencies of integration and support systems such as C2, communications, computers, system integration (SI), space-based, airborne, ground-based and seaborne ISR, IW and an integrated logistics support system (ILSS). When comparing defence forces at the strategic level, a country may like to evaluate the RMA capability of its forces with respect to the RMA capabilities of its adversaries. The methodology described below can be used to carry out such an evaluation, as will be shown in the sections below. The methodology is based on the AHP.

11.2 Components of the RMA Force

The RMA force comprises of four major components – ISR, C4, IW and ILSS. Further ISR is divided into space-based, airborne, seaborne and ground-based ISR. C4 comprises command and control (C2), communications, computers and system integration (SI). For this chapter subdivisions of IW and ILSS are not considered as explicit systems. However, the methodology can be generalised to include these subdivisions as well. The various components and sub-components of the combat support system are shown in Figure 11.1.

We define the revolution in military affairs index (RMAI) or combat support system effectiveness measure (CSSEM) of a force as

$$CSEEM = \left[1 + \sum_{i=1}^{N} W_i V_i\right] \times \left[1 + \sum_{i=1}^{N} S_i V_i\right] \tag{11.1}$$

where W_i is the weight or importance of the ith RMA system, V_i is the capability of the ith RMA system of the force being evaluated, S_i is the combat synergy index

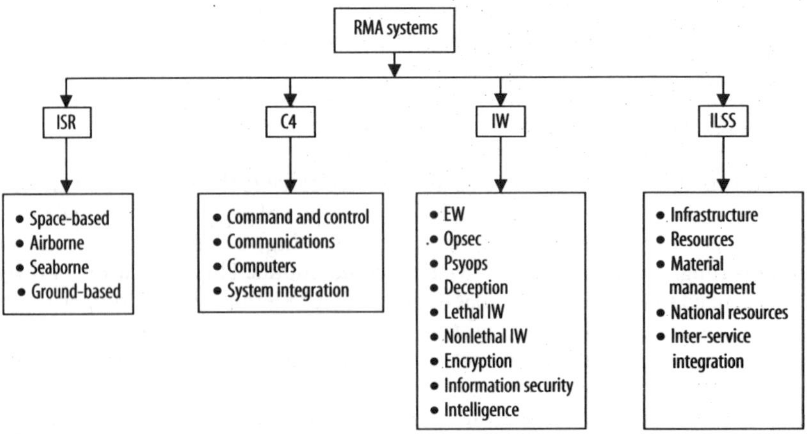

Figure 11.1 RMA systems.

(CSI) of the ith RMA system, which is defined as the degree of dependency of other RMA systems on it. Also,

$$V_i \in [0,1] \text{ and } W_i \in [0,1],\ S_i \in [0,1],\ \sum_{i=1}^{N} S_i = 1 \text{ and } \sum_{i=1}^{N} W_i = 1$$

Also, $\Sigma_{i=1}^{N} W_i V_i$ is defined as the RMA system weighted value or the RMA system importance in combat and $\Sigma_{i=1}^{N} W_i S_i$ is defined as the RMA system synergy value for the force. The procedure for the evaluation of W_i and S_i is described in the next section. The evaluation procedure for V_i, the RMA system capability value of a force, is described in Section 11.4.

11.3 Evaluation of RMA Systems and Integration Indices

The combat support systems effectiveness measure (CSSEM) or RMA index has two components, namely RMA system weights, and combat synergy indices (CSI). In the evaluation of these weights and indices the following steps should be followed.

Step 1: Construct the connectivity/dependency diagram of RMA systems interdependencies (see Figure 11.2).

Step 2: Construct the dependency matrix (A) from the connectivity diagram.

Step 3: Find the principal eigenvalue of A and the corresponding normalised principal right eigenvector (let it be S). The elements of S are the combat synergy indices for the RMA system.

Step 4: Find the principal eigenvalue of A^T and the corresponding normalised right eigenvector. (Let it be D)

Step 5: Take the product of the corresponding elements of S and D; normalise these products and call it W. The elements W_i's are the weights or importance of the ith combat support system in combat dynamics.

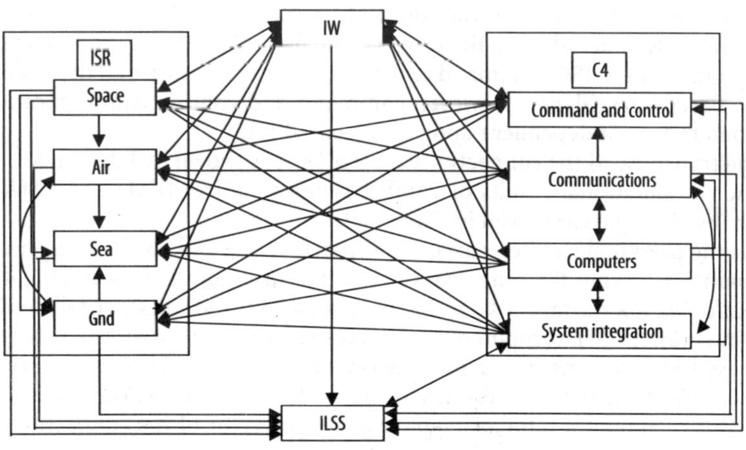

Figure 11.2 Connectivity diagram.

Table 11.1 Connectivity matrix *A*.

		I1	*I2*	*I3*	*I4*	*C1*	*C2*	*C3*	*C4*	*W*	*L*
ISR	Space-based ISR (I1)	1	1	1	1	1	0	0	0	1	1
	Airborne ISR (I2)	0	1	1	1	1	0	0	0	1	1
	Seaborne ISR (I3)	0	0	1	0	1	0	0	0	1	1
	Ground-based ISR (I4)	0	1	1	1	1	0	0	0	1	1
C4	Command and control (C1)	0	0	0	0	1	0	0	0	1	0
	Communications (C2)	1	1	1	1	1	1	1	1	1	1
	Computers (C3)	1	1	1	1	1	1	1	1	1	1
	System integration (C4)	1	1	1	1	1	1	1	1	1	1
	IW (W)	1	1	1	1	1	1	1	1	1	1
	ILSS (L)	0	0	0	0	1	0	0	1	0	1

Table 11.2 Transpose of the connectivity matrix A^T.

		I1	*I2*	*I3*	*I4*	*C1*	*C2*	*C3*	*C4*	*W*	*L*
ISR	Space-based ISR (I1)	1	0	0	0	0	1	1	1	1	0
	Airborne ISR (I2)	1	1	0	1	0	1	1	1	1	0
	Seaborne ISR (I3)	1	1	1	1	0	1	1	1	1	0
	Ground-based ISR (I4)	1	1	0	1	0	1	1	1	1	0
C4	Command and control (C1)	1	1	1	1	1	1	1	1	1	1
	Communications (C2)	0	0	0	0	0	1	1	1	1	0
	Computers (C3)	0	0	0	0	0	1	1	1	1	0
	System integration (C4)	0	0	0	0	0	1	1	1	1	1
	IW (W)	1	1	1	1	1	1	1	1	1	0
	ILSS (L)	1	1	1	1	0	1	1	1	1	1

The connectivity/dependency diagram is shown in Figure 11.2.

The corresponding connectivity matrix (*A*) is shown in Table 11.1.

Each row of the matrix shows the RMA systems which are dependent on the other RMA systems labeling the row; e.g. the first row indicates that space, air, sea and ground-based ISR, command and control, IW and ILSS are all dependent upon space-based surveillance. However, command and control, communications and computers are not dependent upon space-based ISR.

The transpose of the connectivity matrix (A^T) is shown in Table 11.2.

The computational results of combat synergy indices (CSI) and RMA system normalised weights are given in Table 11.3.

S is the principal right eigenvector of matrix *A* and *D* is the right eigenvector of the matrix A^T. From the rank ordering, in the last column of Table 11.3, it is clear that information warfare (IW) is the most important RMA system, followed by system integration (SI), computers, and communications. However, the contribution of these RMA systems to combat synergy are same, i.e. 16.5% each. The synergy index of the RMA system indicates the dependency of other RMA systems on it.

In this methodology, the principal right eigenvector of the connectivity matrix has been chosen as a better indicator of importance and synergy of the combat support system than the simple counting of ones in the rows of the matrix. This is

Table 11.3 RMA system normalised weights and combat synergy indices.

RMA system	*Combat support system dependency indices (DI)*	*Combat synergy indices (SI)*	*Sub-system activity index $A_i = DI \times SI$*	*RMA system normalised weight (W_i)*	*Order of ranking*
Space-based ISR	0.064	0.083	0.00531	0.0598	VIII
Airborne ISR	0.096	0.070	0.672	0.0757	IV
Seaborne ISR	0.115	0.047	0.00541	0.0609	VI
Ground-based ISR	0.096	0.070	0.00672	0.0757	IV
Command and control	0.165	0.033	0.00545	0.0614	V
Communications	0.054	0.165	0.00891	0.1004	III
Computers	0.054	0.165	0.00891	0.1004	III
System integration	0.076	0.165	0.01254	0.1413	II
IW	0.142	0.165	0.2343	0.2639	I
ILSS	0.138	0.039	0.00538	0.0606	VII

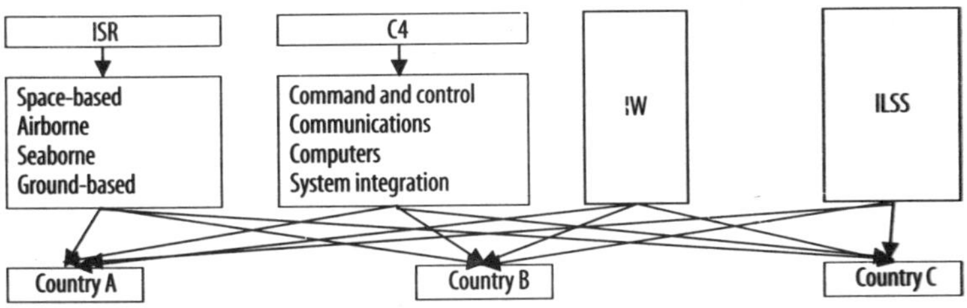

Figure 11.3 Hierarchy for the RMA system evaluation.

because it has been observed that the principal right eigenvector of the matrix reflects not only the direct dependencies but also the indirect dependencies of RMA systems on each other.

Evaluation of the RMA Index

In this section the use of the analytic hierarchy process (AHP) for the evaluation of the RMA capabilities of opposing countries is described. The first step in the AHP is to structure the problem as a hierarchy. The use of the AHP for the RMA capability evaluation is illustrated using the following example. Consider three countries A, B and C whose RMA capabilities are to be assessed. The hierarchy for the RMA system evaluation is shown in Figure 11.3.

The capabilities of the three countries being compared are assimilated into pairwise reciprocal matrices on the basis of expert opinion. While comparing these capabilities, the experts may consider the factors shown in Table 11.4 for each of the RMA systems.

The pairwise comparisons for each RMA system component are obtained in reciprocal matrices as shown below. The normalised right eigenvector corresponding to the principal eigenvalue of these matrices gives the relative scaling of the RMA system capability of each of the three countries.

Space-based ISR

	A	B	C	Eigenvector
A	1	2	3	0.54
B	1/2	1	2	0.297
C	1/3	1/2	1	0.163

Airborne ISR

	A	B	C	Eigenvector
A	1	1/5	1/3	0.105
B	5	1	3	0.638
C	3	1/3	1	0.258

Seaborne ISR

	A	B	C	Eigenvector
A	1	3	1/2	0.309
B	1/3	1	1/5	0.109
C	2	5	1	0.582

Ground-based ISR

	A	B	C	Eigenvector
A	1	1/3	1/3	0.142
B	3	1	1	0.429
C	3	1	1	0.429

Command and control

	A	B	C	Eigenvector
A	1	3	2	0.547
B	1/3	1	2	0.263
C	1/2	1/2	1	0.190

Communications

	A	B	C	Eigenvector
A	1	5	7	0.732
B	1/5	1	3	0.188
C	1/7	1/3	1	0.08

Computers

	A	B	C	Eigenvector
A	1	2	2	0.5
B	1/2	1	1	0.25
C	1/2	1	1	0.25

System integration

	A	B	C	Eigenvector
A	1	5	2	0.607
B	1/5	1	3	0.238
C	1/2	1/3	1	0.155

IW

	A	B	C	Eigenvector
A	1	5	3	0.638
B	1/5	1	5	0.258
C	1/3	1/5	1	0.104

ILSS

	A	B	C	Eigenvector
A	1	2	2	0.493
B	1/2	1	2	0.311
C	1/2	1/2	1	0.196

Table 11.4 Factors to be considered for RMA systems of opposing countries.

	RMA system	*Factors*
ISR	Space-based ISR	Communications satellites, navigation satellites, meteorological satellites, imaging reconnaissance satellites, relay satellites
	Airborne ISR	AWACS/AEW, air-based ground tactical reconnaissance, maritime reconnaissance, airborne ground strategic reconnaissance, airborne battlefield surveillance capability (UAV/RPV capability)
	Seaborne ISR	Surface tactical recce, sub-surface tactical recce, strategic recce, coast-based strategic recce
	Ground-based ISR	Tactical recce, air space recce, strategic recce
C4	Command and control	Strategic C2, tactical C2, politics military C2, inter-services integration
	Communications	Utilisation of EM spectrum, communication system, communication security, merger with national telecom system
	Computers	Computing power, software capability, knowledge processing and automated decision-making capability
	System integration	Data fusion, national level interoperability, inter-services operability, integration of ISR, C2, ILSS with lethal system, fault tolerant communication/computing, interoperability standards
Information warfare (IW)		Operations security, psychological operations, EW, deception lethal IW, non-lethal IW, cryptology, computer security, info security, intelligence
ILSS		Armed forces resources, logistics infrastructure, material management, national resources, inter-service integration

Table 11.5 Values of the *i*th RMA system component for each of three countries.

RMA system		*Country A*	*Country B*	*Country C*
ISR	Space-based	0.54	0.297	0.163
	Airborne	0.105	0.638	0.258
	Seaborne	0.309	0.109	0.582
	Ground-based	0.142	0.429	0.429
C4	Command and control	0.547	0.263	0.190
	Communications	0.732	0.188	0.08
	Computers	0.5	0.25	0.25
	System integration	0.607	0.238	0.155
IW		0.638	0.258	0.104
ILSS		0.493	0.311	0.196

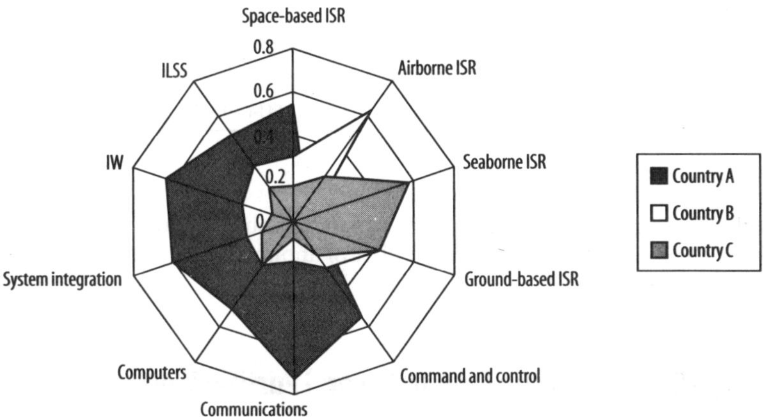

Figure 11.4 RMA capabilities of countries A, B and C.

The corresponding normalised principal right eigenvectors of these matrices gives the values (V_i) of the *i*th RMA system component for each of the countries A, B and C. The values (V_i) as computed using methodology are shown in Table 11.5.

The Kiviat chart in Figure 11.4 shows the comparison of the RMA components of the three countries.

The values (V_i) computed in Table 11.5 and the weights (W_i) and combat synergy indices (S_i) computed as shown in Table 11.3 are used to compute the CSSEMs for countries A, B and C. These computations are shown in Table 11.6.

For Country A:

$$
\begin{aligned}
\text{RMA Index} &= \left[1 + \sum_i W_i V_i\right] \times \left[1 + \sum_i S_i V_i\right] \\
&= (1 + 0.4414) \times (1 + 0.446) \\
&= (1.4414) \times (1.446) \\
&= 2.08426
\end{aligned}
$$

Similar computation shows the RMA Index for countries B and C as 1.3872 and 1.22877 respectively. Using the RMA index ratio between the two countries one can

Table 11.6 CSSEMs for three countries.

CSS		W_i from Table 11.3	S_i from Table 11.3	V_i			W_iV_i			S_iV_i		
				A	B	C	A	B	C	A	B	C
ISR	Space	0.0598	0.083	0.54	0.297	0.163	0.0323	0.0178	0.0097	0.045	0.025	0.014
	Air	0.0757	0.070	0.105	0.638	0.258	0.0079	0.0483	0.0195	0.007	0.045	0.018
	Sea	0.609	0.047	0.309	0.109	0.582	0.188	0.066	0.354	0.015	0.005	0.027
	Ground	0.0757	0.070	0.142	0.429	0.429	0.0107	0.0325	0.0325	0.01	0.03	0.03
C4	Command and control	0.0614	0.033	0.547	0.263	0.190	0.0336	0.0161	0.0117	0.018	0.009	0.006
	Communi-cation	0.1004	0.165	0.732	0.188	0.08	0.0735	0.0189	0.008	0.121	0.031	0.013
	Computers	0.1004	0.165	0.50	0.25	0.25	0.0502	0.0251	0.0251	0.083	0.041	0.041
	System integration	0.1413	0.165	0.607	0.238	0.155	0.0858	0.0336	0.0219	0.1	0.039	0.026
IW		0.2639	0.165	0.638	0.258	0.104	0.1684	0.0681	0.0274	0.105	0.043	0.017
ILSS		0.0606	0.039	0.493	0.311	0.196	0.0299	0.0188	0.0119	0.019	0.012	0.008
Column sum							0.4414	0.1806	0.106	0.446	0.175	0.111

infer that Country A has nearly 1.7 times better combat support than does Country B. The RMA index can be used in a number of applications, as described in the next section.

11.4 Applications of the RMA Index and Further Work

The RMA index as computed using Equation (11.1) can be used for the evaluation of the force strength (FS) of a force which is defined as the summation of the product of weapon power scores (WPS) as computed in Chapter 10 and the quantity of weapon systems in particular categories, i.e.

$$FS = \sum_{i=1}^{N} (WPS)_i \times (n)_i \tag{11.2}$$

where $(WPS)_i$ is the weapon power score of the ith weapon system ($i = 1, 2, \ldots, N$), (N is the number of categories of weapon system) and $(n)_i$ is the number of ith-category weapon systems available with the force.

We define the force potential (FP) of a force as

$$FP = FS \times \text{RMA index} \tag{11.3}$$

where FS is the force strength of the force, and the RMA index is the combat support system effectiveness measure. Further combining Equations (11.1) and (11.3), we get

$$FP = (FS) \times \left[1 + \sum_{i=1}^{N} W_i V_i\right] \times \left[1 + \sum_{i=1}^{N} S_i V_i\right] \tag{11.4}$$

This definition of force potential can be used in force structure planning, wargaming, threat assessment and combat simulations. This will give a more realistic picture of combat capability of the military force, as it incorporates not only the lethal system of the force but also the RMA capabilities as well.

References

1. Jaiswal NK (1997) Military Operations Research: Quantitative Decision-making. Kluwer, Dordrecht.
2. Saaty TL (1980) The Analytic Hierarchy Process. McGraw-Hill, New York.
3. Ince NA, Evrendilek C, Wilhelmsen D, Gezer FI (1997) Planning and Architectural Design of Modern Command, Control, Communications and Information Systems. Kluwer, Dordrecht.
4. Charles M, Garret R (1996) Simulation of C4ISR: Command, control, communications, computers, intelligence, surveillance and reconnaissance. PHALANX, The Bulletin of Military Operations Research 29(1).
5. Davis CC A Methodology for Evaluating and Enhancing C4I Networks. MS Thesis, Air Force Institute of Technology, Wright-Patterson AFB, OH.
6. McKethan C (1997) US C4I and Logistics Vulnerability to Offensive Information Warfare. Naval War College, 13 June.

12 Transition to Nuclear War

12.1 Wars and Nuclear Weapons

Despite their horrifying destructive power, seen at Hiroshima and Nagasaki, the world has not freed itself from nuclear weapons. Nuclear weapons add a radically different dimension to the war-making capability of a nation. A nation in the modern world of economic, commercial and technological geopolitics influenced by increasing globalisation and the growing covert and overt proliferation of weapons of mass destruction (WMD) cannot remain uninfluenced by these trends and events. Applying force as a means of achieving political objectives has been followed constantly throughout the centuries by powerful nations. A major concern today when applying force to subjugate an opponent to the political will of the nation remains the question of what will happen if the other side resorts to nuclear war. This question always lingers because nuclear weapons have immense destructive power. Although it is clear that the use of nuclear weapons cannot help any nation, but rather will create many problems for it, their possession has nevertheless been advocated [2] as a deterrent to achieve political goals.

When planning for a military solution to a political problem in a scenario where opponents possess nuclear weapons it is of major importance to assess when a particular situation may be transformed into war. The transition from peace to tension to actual hostilities between two nations is at one level, which may not have a major impact on the world situation. However, the transition of a war situation to a nuclear war or even a "one-off" nuclear incident can cause major concern the world over. Hence when planning for any impact of a hostile situation anywhere in the world, it is of paramount importance to evaluate when events may escalate to nuclear warfare. The analysis of the factors that can push a situation to nuclear flashpoint is also important if these factors are to be controlled and disaster avoided.

Nuclear weapons have immense destructive power [1]. There are many questions to be answered when planning for any political and military flareup that has a chance of exploding into an exchange of nuclear weapons. Questions such as who in specific scenarios will initiate a nuclear threat, a nuclear posture and a nuclear war are difficult to answer. Another question may be whether either opponent has enough nuclear weapons to think if they can gain a military or political advantage by using them or whether the nuclear weapons are there just to deter its adversary from gaining advantage in conventional warfare. Other factors to be taken into account when analysing such situations may be the perceived casualties and the extent of the damage to be inflicted on the enemy, and also the expected damage from a second strike. Another challenge is whether the doctrine of no first use (NFU) will prove to be non-practical and whether developing sufficient capability to absorb a first strike and retaliate with sufficient power in the second one will be effective.

In trying to answer these strategic questions, it is pertinent to understand how wars can happen, how hostilities between nations can reach such a point that opponents have to resort to military force, and how these factors can lead to an actual exchange of nuclear weapons. It is also important to juxtapose the various stages in the geopolitical, economic and ideological relations between two nations when one or both of them possess nuclear weapons capabilities and delivery systems. The increasing proliferation of such weapons leads to capabilities that are difficult to rationalise as purely defence mechanisms. Since it is unlikely that in the near future we will have a world free of nuclear weapons, the point at which a war can become a nuclear one becomes a critical input to the strategic decision-making of nations preparing to defend themselves or achieve their political goals through the use of military force.

12.2 Various Scenarios and Stages Leading to Nuclear Wars

At the conceptual level, two nations can become mutually hostile for a variety of reasons that include ideology, economy, resource control or political systems. The conceptual framework for the study of such interactions between two nations starts with what each country considers as its national interests. When these somehow become entangled, antagonistic relations follow. These can become more and more hostile if the conflict is not resolved amicably within a certain time frame. Also, if the perceived relative military capabilities of the nations are asymmetrical, the country that sees itself as more powerful has an incentive to use its military power to resolve the hostile situation so as to meet its national interests. The deterrent scope of nuclear weapons is an incentive for every nation to protect itself from such a scenario.

The road to nuclear war requires that at least one of the actors should possess the capability to strike with nuclear weapons. The nuclear capability has three critical dimensions. The first is nuclear warheads – the capability to produce a chain reaction in a fissile material such as uranium or plutonium requires a considerable research and development effort. Not many countries possess this capability. However, there are indications that nuclear weapons have proliferated from nuclear "have" nations to others that were nuclear "have-nots" [3]. The second dimension of nuclear weapon capability is a delivery system. The warheads need to be delivered on target using one of the so-called "triad" of specialised delivery systems, which may be long-range bomber aircraft, land-based missile systems or submarine-based missile systems. The third critical dimension of nuclear war-making capability is the C4ISR system. The command, control, communications, computers, intelligence, surveillance and reconnaissance (C4ISR) system is an integrated part of modern military force. However, for nuclear weapons the C4ISR is critically important, as it is linked to civilian authority as well. Here, we assume that the government of the nation is a democratically elected civilian one and controls the nuclear weapons. However, the execution and launch of nuclear weapons has to be routed through the military forces, as they have the wherewithal to execute the launch. Herein lies the danger of a command and control breakdown, which can lead to the inadvertent launch of nuclear weapons during a crisis situation.

Another important parameter that can effect the transition concerns the perception of each nation about whether the other has a nuclear capability or not. Such perceptions may produce a fog of war and impact rational decision-making during a crisis. From the point of view of possession and perception, we thus have 16 scenarios as shown in Table 12.1. "Asymmetrical possession" implies that only one side has the capability to conduct nuclear war, while "symmetrical possession" implies that both sides have it. "No capability" implies that none of the adversaries has nuclear war-making capability. If both sides perceive wrongly about each other this is termed "symmetrical incorrect perceptions". One side perceiving wrongly about the other, while the other side perceives correctly, is "asymmetrical incorrect perceptions". If both sides perceive correctly about each other, we have "symmetrical correct perceptions".

In the no-possession scenarios (shown with a dark background), there is no possibility of nuclear war. In every other scenario nuclear war can take place, though with a probability which varies in each case. Any adversary in any of these scenarios needs to evaluate whether or not the hostilities will lead from peace or a conventional war scenario to a nuclear war. A rational escalatory framework for moving up to nuclear war, not necessarily in the sequential manner, can be articulated in the following phases:

Peace or status quo phase: There are no overt military hostilities between adversaries.

Threat or warning phase: At least one of the adversaries resorts to giving warnings to the other orally or through actions such as troop movements on the border,

Table 12.1 Nuclear war scenarios based on possession and perception.

Country A possesses	Perceives B possesses	Symmetrical incorrect perceptions Asymmetrical possession	Asymmetrical incorrect perceptions Asymmetrical possession	Asymmetrical incorrect perceptions Symmetrical possession	Symmetrical correct perceptions Symmetrical possession
	Perceives B does not possess	Asymmetrical incorrect perceptions Asymmetrical possession	Symmetrical correct perceptions Asymmetrical possession	Symmetrical incorrect perceptions Symmetrical possession	Asymmetrical incorrect perception Symmetrical possession
Country A does not possess	Perceives B possesses	Asymmetrical incorrect perceptions Asymmetrical possession	Symmetrical incorrect perceptions Asymmetrical possession	Symmetrical correct perceptions Asymmetrical possession	Asymmetrical incorrect perception Asymmetrical possession
	Perceives B does not possess	Symmetrical correct perceptions No possession	Asymmetrical incorrect perceptions No possession	Asymmetrical incorrect perception Asymmetrical possession	Symmetrical incorrect perception Asymmetrical possession
	PERCEPTIONS	Perceives A does not possess	Perceives A possesses	Perceives A does not possess	Perceives A possesses
POSSESSION		Country B does not possess		Country B possesses	

trade or economic sanctions etc. Actions in this phase may also include inciting insurgencies and uprisings in the adversary's occupied territories or fomenting religious, ethnic or cultural divisions in the adversary's population.

Limited conventional operations: Actual military operations take place. These operations are however limited to certain sectors and involve minimum forces. The objectives in this phase are to create disturbances across the border or take advantage of tactical positions. These operations may be conducted through irregular forces in the form of low-intensity conflicts (LIC).

Conventional military operations: Regular war; the whole conventional military might of each adversary is involved in mid-intensity conflict.

Nuclear threat or warning: This phase involves a warning of the use of nuclear weapons in conventional war situations by either adversary or both. This phase varies as per the scenarios depicted in Table 12.1 and is actually based on the perceptions and the known or proven nuclear capabilities of the adversaries.

Nuclear war: The final escalatory phase, which should be avoided as far as possible. Rationality should not be lost during this phase as it can lead to disastrous results for all parties involved. However, it is of paramount importance to study parameters that can affect this tilt towards irrationality.

In the next section we look at the parameters reflecting the transition to nuclear war and how the AHP can be used to study various possibilities.

12.3 Factors Affecting the Transition To Nuclear War

We consider a hypothetical situation in which Country A has to evaluate whether and when its adversary, which possesses nuclear weapons, will resort to a nuclear strike or strikes if Country A starts conventional military operations to resolve the conflict. In the scenario depicted the following alternatives are debated:

- *A1–Immediate strike*: The adversary will resort to a nuclear strike immediately (within 1 week).
- *A2–Deferred strike*: The adversary will wait till it suffers heavy losses in 50% of all sectors (estimated time: 4–5 weeks).
- *A3–Last-resort strike*: The adversary will wait till it is about to be comprehensively defeated (8–10 weeks).
- *A4–No strike*: The adversary will not resort to nuclear weapons and will seek international pressure on Country A.

The military and civilian decision-makers of Country A have to evaluate these options for the adversary and then decide on their conventional operations objectives and strategy. Their main aim is for the adversary not to resort to nuclear strikes at all. The hierarchy of parameters that affect this decision is shown in Figure 12.1. These parameters are defined as follows:

Conventional military balance (CMB): This is an important parameter, as the success of military operations will depend upon the conventional military capability differential that Country A has over its adversary. If this is too much then the adversary is likely to resort to a nuclear threat and in turn a nuclear strike earlier during the escalatory phase of the war.

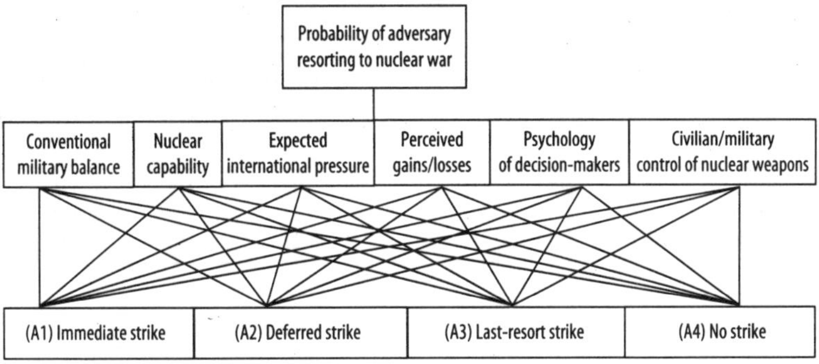

Figure 12.1 Hierarchy for evaluating transitions to nuclear war.

Nuclear capability (NC): This parameter reflects the ability of the adversary to conduct nuclear operations in the given time frame. It also depends upon the state in which the adversary has kept its nuclear force. This state can be: (1) just the capability, (2) recessed deterrent, or (3) ready arsenal. Depending on the current state of the adversary's nuclear capability, the following four alternatives can be evaluated.

Expected international pressure (EIP): If Country A makes rapid progress in conventional military operations, the adversary will expect international pressure on Country A to mount. If that doesn't happen, the adversary will resort to nuclear blackmail or threaten to use nuclear weapons. Actual use of nuclear weapons is also possible.

Perceived gains/losses (PG/L): If the adversary perceives the status at a particular instance of time as a major loss for it, then it has more incentive to resort to a nuclear strike.

Psychology of decision-makers (PDM): The complexity of the decision-making process and the psychology of the key decision-makers of the adversary will impact the decision to use nuclear weapons to a great extent.

Civilian–military control of nuclear weapons (CMC): The complicated military–civilian command and control structures and dynamics can play a major role in escalating the situation to a nuclear flashpoint. This is true especially when there is mistrust between the civilian and military organizations of the adversary.

Table 12.2 gives the first-level matrix for evaluating relative weights of the parameters affecting the alternatives.

Tables 12.3 to 12.8 show the matrices at the alternatives level to be compared by the decision-makers of Country A with respect to first-level criteria.

The computations by combining the weights and ratings obtained in the above tables are shown in Table 12.9. As can be seen, it is evident that the adversary would resort to a deferred strike (41.9%) while it can launch an immediate nuclear strike as well (18.6%). The probability that the adversary would not resort to a nuclear strike at all is 26.6%, which is quite low. Hence Country A is deterred by the chances that the adversary will resort to a nuclear strike if Country A starts conventional military operations.

From the Kiviat charts for all the four alternatives shown in Figures 12.2 to 12.5, one can see the contribution of various parameters to the final rating.

Table 12.2 Comparison matrix for first-level criteria.

	CMB	*NC*	*EIP*	*PG/L*	*PDM*	*CMC*	*NEV*
CMB	1	1/3	1/2	2	1/3	2	0.119
NC	3	1	2	2	1/2	3	0.230
EIP	2	1/2	1	3	1/3	1/2	0.109
PG/L	1/2	1/2	1/3	1	1/4	1/3	0.056
PDM	3	2	3	4	1	2	0.283
CMC	1/2	3	2	3	1/2	1	0.204

Table 12.3 Comparison matrix with respect to conventional military balance.

	A1	*A2*	*A3*	*A4*	*NEV*
A1 Immediate strike	1	1/3	2	1/3	0.137
A2 Deferred strike	3	1	5	1	0.394
A3 Last-resort strike	1/2	1/5	1	1/5	0.075
A4 No strike	3	1	5	1	0.394

Table 12.4 Comparison matrix with respect to nuclear capability.

	A1	*A2*	*A3*	*A4*	*NEV*
A1 Immediate strike	1	1/3	7	1/4	0.187
A2 Deferred strike	3	1	5	1/2	0.310
A3 Last-resort strike	1/7	1/5	1	1/3	0.065
A4 No strike	4	2	3	1	0.437

Table 12.5 Comparison matrix with respect to expected international pressure.

	A1	*A2*	*A3*	*A4*	*NEV*
A1 Immediate strike	1	1/5	1/7	1/9	0.037
A2 Deferred strike	5	1	1/3	1/7	0.105
A3 Last-resort strike	7	3	1	1/5	0.213
A4 No strike	9	7	5	1	0.645

Table 12.6 Comparison matrix with respect to perceived gains/losses.

	A1	*A2*	*A3*	*A4*	*NEV*
A1 Immediate strike	1	3	7	1/3	0.279
A2 Deferred strike	1/3	1	3	1/5	0.113
A3 Last-resort strike	1/7	1/3	1	1/7	0.050
A4 No strike	3	5	7	1	0.558

Table 12.7 Comparison matrix with respect to psychology of decision-makers.

	A1	*A2*	*A3*	*A4*	*NEV*
A1 Immediate strike	1	1/3	1	7	0.200
A2 Deferred strike	3	1	5	9	0.580
A3 Last-resort strike	1	1/5	1	7	0.182
A4 No strike	1/7	1/9	1/7	1	0.037

Table 12.8 Comparison matrix with respect to civilian/military control of nuclear weapons.

	A1	*A2*	*A3*	*A4*	*NEV*
A1 Immediate strike	1	1/3	3	7	0.246
A2 Deferred strike	3	1	7	9	0.584
A3 Last-resort strike	1/3	1/7	1	9	0.137
A4 No strike	1/7	1/9	1/9	1	0.034

Table 12.9 Final rating of adversary's likely actions.

Parameter	*Weight*	*A1*	*A2*	*A3*	*A4*
Conventional military balance	0.119	0.137	0.394	0.075	0.394
Nuke capability	0.230	0.187	0.310	0.065	0.437
Expected international pressure	0.109	0.037	0.105	0.213	0.645
Perceived gains/losses	0.056	0.279	0.113	0.050	0.558
Psychology of decision-makers	0.283	0.200	0.580	0.182	0.037
Civilian/military control of nuclear weapons	0.204	0.246	0.584	0.137	0.034
Final rating		0.186	0.419	0.129	0.266

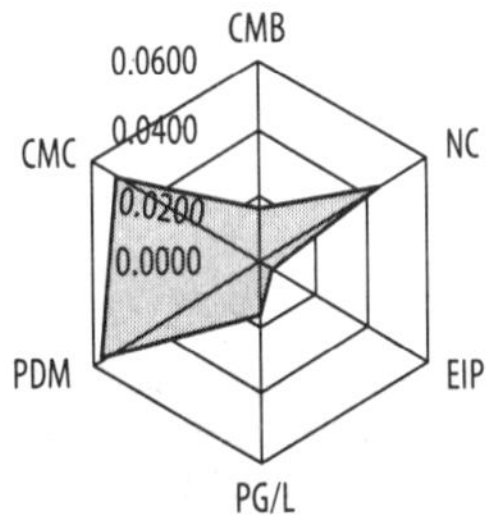

Figure 12.2 Immediate strike (A1).

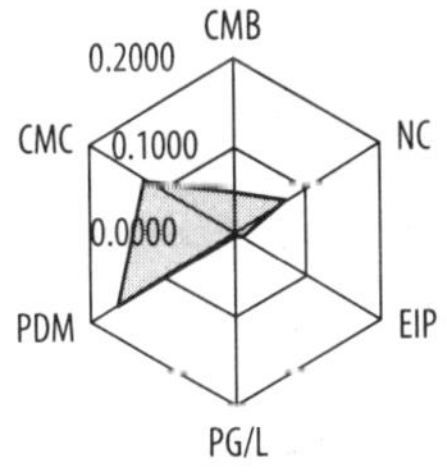

Figure 12.3 Deferred strike (A2).

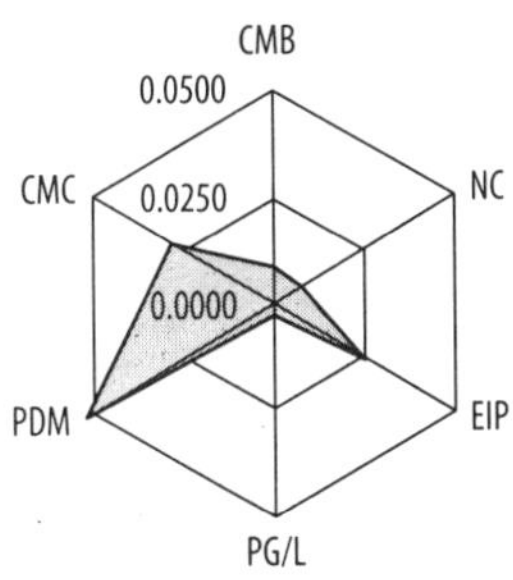

Figure 12.4 Last-resort strike (A3).

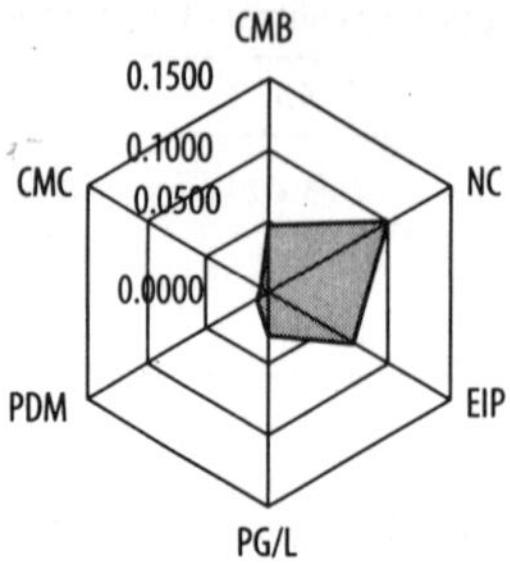

Figure 12.5 No strike (A4).

It is evident from these charts that as per the perception of decision-makers of Country A, the adversary will resort to a deferred nuclear strike. This is a crucial input to any war strategy for Country A to execute.

Such a framework, analysing transitions to high-intensity conflicts involving nuclear weapons, can be used to answer a variety of strategic questions.

References

1. Tsipis K (1983) Understanding Nuclear Weapons. Wildwood House, London.
2. Spulak RG (1997) The case in favor of US nuclear weapons. PARAMETERS (Spring).
3. Lee R (2000) Nuclear smuggling: patterns and responses. PARAMETERS (Spring).

Index

M

N

O

P

Q

R

S

T

U

V

W